Berries
COOKBOOK

by Carol Ann Shipman

hancock
house

ISBN 0-88839-512-4
ISBN 0-88839-582-5 (Alaskan edition)
Copyright © 2004 Carol Ann Shipman

Cataloging in Publication Data

Shipman, Carol Ann, 1944–
    Berries cookbook / Carol Ann Shipman.

(Nature's gourmet series)
Includes index.
ISBN 0-88839-512-4 — ISBN 0-88839-582-5 (Alaskan ed.)

1. Cookery (Berries) I. Title. II. Series.
TX813.B4S54 2004   641.6'47   C2003-910989-5

All rights reserved. No part of this publication may be reproduced, stored in a retrieval system or transmitted, in any form or by any means, electronic, mechanical, photocopying, recording, or otherwise, without the prior written permission of Hancock House Publishers.
Printed in China—JADE

Editing: Nancy Miller, Yvonne Lund
Series design and production: Nando DeGirolamo
Photographic sources listed on page 92.

Published simultaneously in Canada and the United States by

hancock house

HANCOCK HOUSE PUBLISHERS LTD.
19313 Zero Avenue, Surrey, B.C. V3S 9R9
(604) 538-1114 Fax (604) 538-2262

HANCOCK HOUSE PUBLISHERS
1431 Harrison Avenue, Blaine, WA 98230-5005
(604) 538-1114 Fax (604) 538-2262
Web Site: www.hancockhouse.com *email:* sales@hancockhouse.com

dedication

This book is dedicated to my son, John Thomas Wood, with love.

acknowledgments

I was very fortunate to work with a wonderful team to produce this book. I appreciate and thank all the people who were involved, and give special thanks to Richard Shipman, my husband, for the constant support and understanding testing endless recipes for all of the cookbooks.

To my publisher, David Hancock, who shared my vision for the series, thank you for your patience and enthusiasm for this book.

A special thank you to Richard Groenheyde, production coordinator, for his talent to meet deadlines and fine tune the changes before printing.

To Nando DeGirolamo, my designer and partner, for his outstanding design for the entire series, that makes this book on first printing a best seller.

Special thanks go to Ginnine Kowalchuk for her extraordinary help and creative ideas with the first draft.

To my dear friend and catering department head, Eva Toni, who was always at my side during the creation of recipes for this book. Thank you for always caring and asking that question, "How's the book coming?"

Thank You!

contents

6 blueberries

34 strawberries

blackberries 64

saskatoon berries 80

blueberries

Blueberries can change color when cooked. Acids, such as lemon juice and vinegar, cause the blue pigment in the berries to turn reddish. Blueberries also contain a yellow pigment which, in an alkaline environment (such as a batter with too much baking soda) can give you greenish blue berries.

To reduce the amount of color streaking, stir your blueberries (right from your freezer, if frozen) into your cake or muffin batter last.

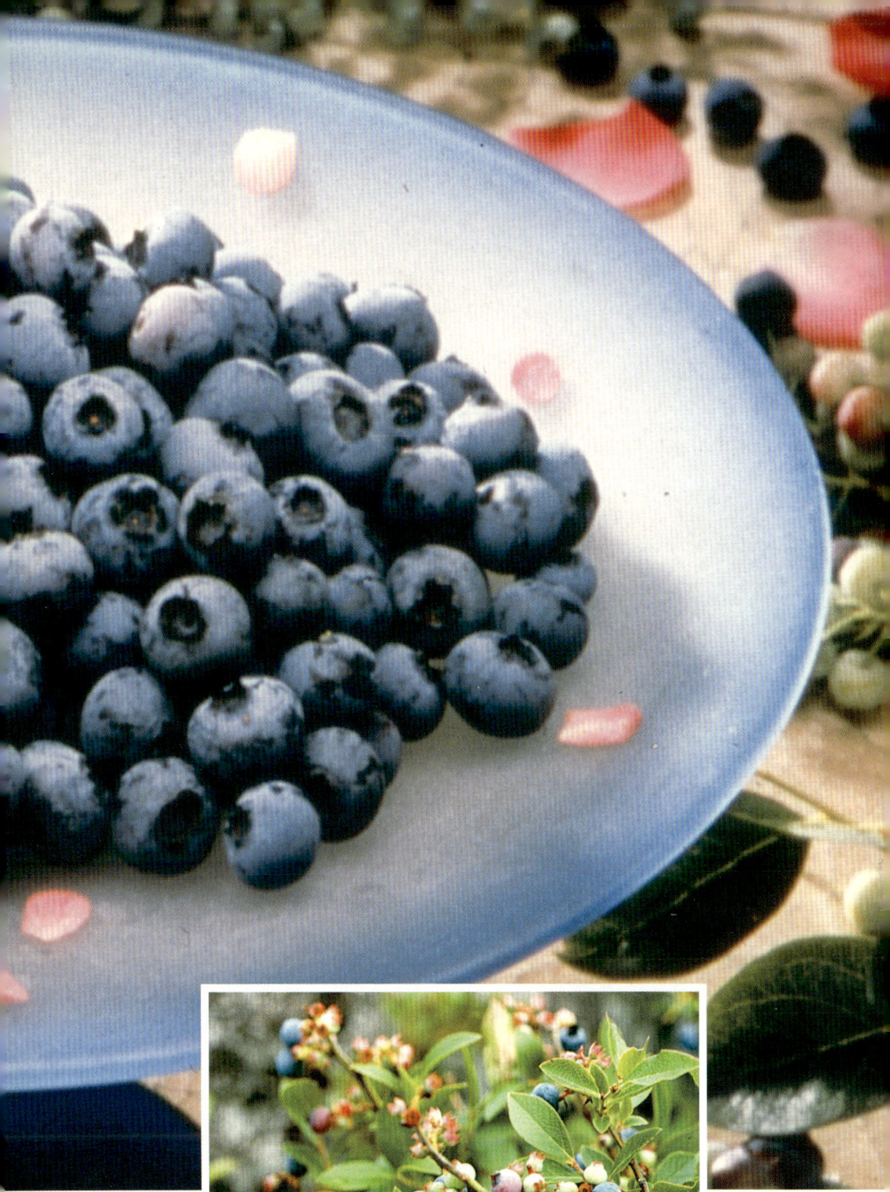

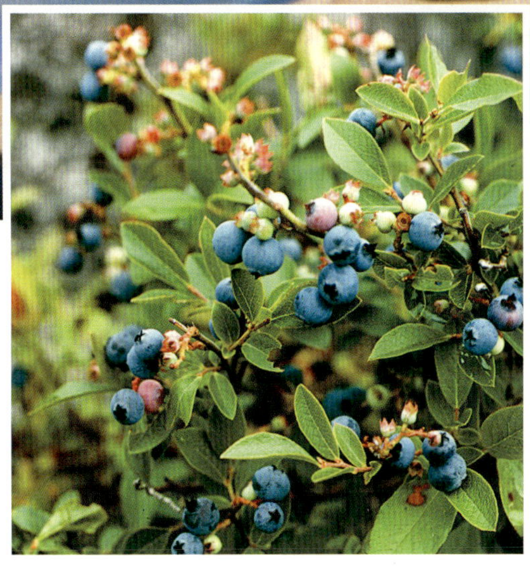

Blueberry Smoothie

BLUEBERRIES

Berry Sparkler

Syrup
In a small saucepan, combine blueberries, lime juice and sugar. Stir over medium heat until sugar dissolves.

Increase heat, bring to a boil and boil for 2 minutes.

Remove from heat and cool.

Each drink
Pour 1/4 cup (60 mL) of syrup over ice cubes in a tall glass. Top with club soda. Stir. Garnish with a lime slice.

Alcoholic drink
Add 1-1/2 ounces (45 mL) of gin or vodka before topping with soda.

SERVES 2

1 cup	**blueberries** fresh or frozen	250 mL
1/2 cup	**lime juice**	125 mL
3/4 cup	**sugar**	175 mL
	club soda	
6	**lime slices**	6

Blueberry Smoothie

In a container of an electric blender combine blueberries, juice, yogurt and sugar. Blend until smooth, about 1 minute.

Serve immediately in tall glasses.

Garnish with blueberry skewers spiraled with thin strips of orange peel if desired.

SERVES 2

2 cups	**blueberries** frozen, slightly thawed	500 mL
1 cup	**pineapple and orange juice blend**	250 mL
8 oz	**vanilla yogurt** low-fat	250 g
2 tsp	**sugar**	10 mL

Blue Witch's Brew

In blender combine blueberries, apple juice, ice cream, milk and cinnamon until smooth.

Serve immediately.

SERVES 3 – 4

2-1/2 cups	**blueberries** fresh	625 mL
1-1/4 cups	**apple juice**	300 mL
1 cup	**vanilla ice cream**	250 mL
1/4 cup	**milk**	60 mL
3/4 tsp	**cinnamon**	4 mL

Carol Ann Shipman **Berries Cookbook**

Blueberry-stuffed French Toast

SERVES 9

12	thin slices French bread 6 cut up into cubes	12
16 oz	cream cheese	500 g
1-1/4 cups	blueberries	300 mL
1 tbsp	butter	15 mL
12	eggs	12
1/3 cup	maple syrup	75 mL
1-1/2 cups	milk	375 mL
1/2 cup	whipping cream	125 mL
1 tbsp	cinnamon/nutmeg blend	15 mL

SAUCE

1 cup	sugar	250 mL
2 tbsp	cornstarch	30 mL
1 cup	water	250 mL
1 tbsp	butter	15 mL
1 cup	blueberries	150 mL

Arrange bread slices in a buttered 9 x 13 inch (22 x 33 cm) pan; cut the cream cheese into squares and place over the bread. Sprinkle the blueberries over the cheese. Arrange the bread cubes over the top of the blueberries – whole bread on the bottom, cube bread on top of this layer.

In a large bowl mix the eggs, syrup, milk, cream, cinnamon and nutmeg. Whisk well, pour over the bread mixture. Chill overnight.

Bake at 350°F (175°C) for about an hour or until it is puffed and golden brown.

Sauce

In a small saucepan, mix the water, sugar, butter and cornstarch over moderately high heat until it is thickened, then add the blueberries. Simmer ten minutes or until blueberries burst, then pour over French toast.

Blueberry-stuffed French Toast

BLUEBERRIES

Ashley Inn Lemon Curd Waffles with Blueberry Sauce

Waffle Batter
Combine waffle mix, milk, water, eggs, vanilla and orange extract.

Lemon Curd
Whisk eggs, sugar and lemon juice over low heat until thickened, stirring constantly. Fold in cream cheese.

Blueberry Sauce
Dissolve sugar in orange juice or water over medium heat. Add berries and stir. Add cornstarch and bring to a boil.

Prepare waffles. Spread 2 tablespoons (30 mL) lemon filling on waffles, fold over and top with blueberry sauce.

SERVES 4

WAFFLE BATTER

4 cups	waffle mix	1 kg
1-1/2 cups	milk	375 mL
1-1/2 cups	water	375 mL
3	eggs	3
2 tsp	vanilla extract	10 mL
2 tsp	orange extract	10 mL

LEMON CURD

3	eggs	3
1/2 cup	sugar	125 mL
1/4 cup	lemon juice	60 mL
8 oz	cream cheese	250 g

BLUEBERRY SAUCE

1 cup	sugar	250 mL
2 tbsp	orange juice	30 mL
3 cups	blueberries	750 mL
1 tbsp	cornstarch	15 mL

Blueberry Fruit Shake

In the container of a food processor or blender, place blueberries, mixed fruit, milk, sugar and vanilla extract; whirl until smooth.

Serve immediately.

YIELD 2 1/2 cups (2 – 3 servings)

2 cups	frozen blueberries	500 mL
1 cup	frozen mixed fruit including cantaloupe, honeydew, grapes, peaches	250 mL
1 cup	milk	250 mL
1 tbsp	sugar	15 mL
2 tsp	vanilla extract	10 mL

Blueberry Ginger Sauce

In a large saucepan, combine ingredients.

Stir in 1/3 cup (75 mL) of water. Over medium-high heat, bring to a boil; cook and stir until sauce thickens, about 1 minute.

Serve with Fluffy Blueberry Pancakes.

YIELD 1 1/2 cups

2 cups	fresh blueberries	500 mL
1/4 cup	sugar	60 mL
1 tbsp	ginger finely chopped crystallized or 1/2 teaspoon (2 mL) dried ground ginger	15 mL

Carol Ann Shipman **Berries Cookbook**

Blueberry Pancakes

BLUEBERRIES

Fluffy Blueberry Pancakes

In a medium-sized bowl, combine ricotta cheese, butter and egg yolks until blended. In a small bowl, stir together flour, sugar and lemon peel. Stir dry ingredients into ricotta mixture. In a medium-sized bowl, beat egg whites until they form soft peaks. Fold egg whites and then blueberries into batter. Over medium heat, form pancakes by spooning 1/4 cup (60 mL) of batter per pancake onto a hot, lightly greased griddle or skillet. Cook pancakes until browned, turning once.

Serve with Blueberry Ginger Sauce.

SERVES 6

1-1/2 cups	**ricotta cheese**	375 mL
	or drained small-curd cottage cheese	
1/4 cup	**melted butter**	60 mL
4	**egg yolks**	4
1/2 cup	**flour**	125 mL
1/4 cup	**sugar**	60 mL
2 tsp	**grated lemon peel**	10 mL
	yellow part only	
8	**egg whites**	8
2 cups	**blueberries**	500 mL
	fresh	

Magnolia Inn French Toast

Spread out the bread slices in a non-reactive dish. In a medium bowl, mix eggs, ice cream, orange juice, vanilla, cinnamon and nutmeg. Pour over bread slices and set aside until you are ready. Can be left overnight covered in the refrigerator. Heat your griddle, melt a small amount of butter and add the soaked bread. Cook until brown on one side, turn and cook about 2 minutes more. Transfer to plates and cover with berries of your choice.

To serve
Heat blueberry and maple syrup together, top with lemon zest.

Variation
Cook apple slices with small amount of butter and brown sugar.

SERVES 6 – 8

1-2	**medium French**	1-2
	baguettes or loaves	
	ends discarded, cut diagonally in 3/4-inch slices	
4	**large eggs**	4
1/2 cup	**premium vanilla**	125 mL
	ice cream	
1/4 cup	**fresh orange juice**	60 mL
1 tbsp	**pure vanilla extract**	15 mL
1 tbsp	**cinnamon**	15 mL
Pinch	**nutmeg**	
	blueberry and pure maple syrup	

BLUEBERRIES

Night Swan Blueberry Cheese Strata

SERVES 4 – 6

6 cups	**bread** cubed, divided	1500 mL
2 cups	**blueberries** fresh or frozen	500 mL
1/3 cup	**sugar**	75 mL
1 cup	**swiss cheese** shredded low-fat	250 mL
4	**eggs** beaten	4
2 cups	**low-fat milk**	500 mL
1 tsp	**vanilla extract**	5 mL
1/4 tsp	**salt**	1 mL
1/2 tsp	**cinnamon**	2 mL

Layer half of the bread cubes, blueberries, sugar, cheese and remaining bread cubes in buttered 8 x 8 inch (20 x 20 cm) baking dish. Combine eggs, milk, vanilla and salt; mix well. Pour mixture over bread; sprinkle with cinnamon.

Cover and refrigerate 6 to 24 hours. Uncover and bake at 325°F (160°C) for 75 minutes or until knife blade inserted near center comes out clean. Let stand 10 to 15 minutes before serving. Top with yogurt or sour cream, or blueberry syrup.

Wilted Spinach with BC Blueberries

SERVES 2 – 3

1 cup	**blueberries** fresh or frozen	250 mL
2 tbsp	**sugar**	30 mL
4	**slices of bacon**	4
2	**shallots** minced	2
2	**garlic cloves** minced	2
1/2 cup	**corn oil**	125 mL
2 tbsp	**balsamic vinegar**	30 mL
2 tbsp	**red wine vinegar**	30 mL
2 tsp	**rosemary** fresh chopped	10 mL
1/2 cup	**toasted hazelnuts** coarsely chopped	125 mL
1	**green apple** peeled, cored and diced	1
3 cup	**spinach** fresh	750 mL
	salt and pepper	

Sprinkle blueberries with sugar. Set aside. Cook bacon in a large heavy skillet over medium heat until crisp. Remove with slotted spoon and set aside. Sauté shallots, and garlic in the bacon fat until soft. Whisk in the oil and vinegars. Stir in rosemary, hazelnuts, apples and blueberries, and cook for one minute. Add the spinach and toss until just wilted, about 30 seconds. Season with salt and pepper to taste. Garnish with crumbled bacon.

Carol Ann Shipman **Berries Cookbook**

BLUEBERRIES

West Coast Blueberry Salad

SERVES 6

Croutons
Preheat the broiler. Brush one side of each baguette slice with olive oil. Place a round of goat cheese on each slice. Sprinkle each round of cheese with fresh thyme. Place the bread on a baking sheet and broil until the edges are golden and the cheese puffs.

Toss the greens with the smoked salmon, red onion, blueberries and blueberry vinaigrette. Divide among six plates and place two croutons on each plate. Serve at once.

Blueberry Vinaigrette
Pour the blueberry vinegar into a small bowl. Whisk in the mustard. In a slow steady stream, whisk the oil into the vinegar. Add the thyme. Season with salt and pepper.

12	**baguette slices** 1/2 inch thick	12
4 oz	**goat cheese log** semi-soft, sliced in 12	125 g
2 tbsp	**olive oil**	30 mL
1/2 tsp	**fresh thyme**	2 mL
12 oz	**mixed baby greens** washed and dried	375 g
4 oz	**dry smoked salmon** crumbled	125 g
1/2 cup	**red onion** thinly sliced	125 mL
1 cup	**blueberries**	250 mL
1/2 cup	**blueberry vinaigrette** (recipe below)	125 mL

BLUEBERRY VINAIGRETTE

1/4 cup	**blueberry vinegar**	60 mL
1-1/2 tsp	**dijon mustard**	7 mL
3/4 cup	**olive oil**	175 mL
1 tsp	**fresh thyme**	5 mL
	salt and fresh ground pepper to taste	

Blueberry Melon Salad

Combine all ingredients except salt.

Season with salt to taste.

Let stand one hour and then refrigerate until serving time.

SERVES 6

2 cups	**blueberries**	500 mL
1 cup	**honeydew melon** diced	250 mL
1/2 cup	**red onion** diced	125 mL
1/4 cup	**cilantro** chopped	60 mL
3 tbsp	**lime juice**	45 mL
1 tbsp	**peanut oil**	15 mL
1 tsp	**lime peel** grated	5 mL
1	**jalapeño pepper** finely chopped	1
	coarse salt to taste	

Carol Ann Shipman **Berries Cookbook**

BLUEBERRIES

Lemon Blueberry & Chicken Salad

SERVES 4

2 cups	**blueberries** fresh or frozen, divided	500 mL
3/4 cup	**lemon yogurt** low-fat	175 mL
3 tbsp	**mayonnaise** reduced-calorie	45 mL
1 tsp	**salt**	5 mL
2 cup	**chicken breasts** cubed cooked	500 mL
1/2 cup	**green onions** sliced	125 mL
3/4 cup	**celery** diagonally sliced	175 mL
1/2 cup	**red pepper** diced	125 mL

Reserve a few blueberries for garnish.

In a medium bowl, combine yogurt, mayonnaise and salt. Add remaining blueberries, the chicken, green onions, celery and red pepper; mix gently. Cover and refrigerate to let flavors blend, at least 30 minutes.

Serve over endive, or other greens, garnished with reserved blueberries and lemon slices.

Lemon Blueberry & Chicken Salad

Blueberry Mediterranean Salad

Combine all ingredients, reserving blueberries to fold in last.

Blueberry Mediterranean Salad

SERVES 10 (1/2-cup servings)		
1 cup	**cherry tomatoes**	250 mL
1 cup	**cucumber** diced	250 mL
1/2	**medium red onion** chopped in 3/4-inch pieces	1/2
1/2	**green pepper** chopped	1/2
1/4 cup	**ripe olives**	60 mL
1/2 cup	**feta cheese** chopped in 1/2-inch pieces	125 mL
3 tbsp	**balsamic vinegar**	45 mL
3 tbsp	**red wine**	45 mL
1 tbsp	**water**	15 mL
3 tbsp	**olive oil**	45 mL
1/2 tsp	**basil** fresh or dried	2 mL
1 cup	**blueberries** fresh or frozen	250 mL

Creamy Smoked Turkey & Blueberry Salad

In a bowl, combine mayonnaise, yogurt, marmalade, lemon juice and pepper. Add peach slices, blueberries and turkey: toss until well coated. Serve on lettuce leaves.

SERVES 8 (1-cup servings)		
1/2 cup	**light mayonnaise**	125 mL
1/2 cup	**yogurt** plain low-fat	125 mL
1/4 cup	**orange marmalade**	60 mL
2 tsp	**lemon juice**	10 mL
1/2 tsp	**ground black pepper**	2 mL
3	**medium peaches** cut in wedges	
2 cups	**blueberries**	500 mL
2 cups	**smoked turkey** cubed	500 mL

Tropical Blueberry, Pineapple & Jalapeño Salad

Combine blueberries, pineapple, rum, jalapeño peppers, lime peel and pepper sauce; mix well.

Serve on a bed of mixed salad greens if desired.

SERVES 12		
2 lbs	**blueberries** fresh or frozen	907 g
2 lbs	**pineapple chunks** fresh or canned	907 g
1/2 cup	**rum** or 2 tbsp rum extract	125 mL
3 tbsp	**jalapeño peppers** chopped fresh	45 mL
2 tbsp	**grated lime peel**	30 mL
1/2 tsp	**hot pepper sauce**	2 mL

Carol Ann Shipman **Berries Cookbook**

BLUEBERRIES

Rabbit Hill Inn Blueberry Burgundy Soup

SERVES 4 – 6		
2 cups	**blueberries**	500 mL
1 cup	**red burgundy wine**	250 mL
4 tbsp	**honey or cornstarch** divided	30 mL
1/4 cup	**orange juice**	60 mL
	dash of cinnamon optional	
1 cup	**vanilla yogurt**	250 mL
1 cup	**sweet white wine**	250 mL
1 cup	**sweet white or red wine**	250 mL

In a pot, cook blueberries, with 1 cup (250 mL) of red wine. Add 2 tablespoons (30 mL) honey, orange juice and cinnamon.

In a bowl, mix 1 cup (250 mL) of white or red wine and 2 tablespoons (30 mL) honey or cornstarch. Stir well and add to soup mixture. Cook for an additional five minutes. Cool. Then add vanilla yogurt and 1 cup (250 mL) of sweet white wine such as a Riesling or sauterne. Serve cold.

Blueberry Orange Soup

SERVES 8		
1-1/4 lbs	**blueberries** fresh or frozen	564 g
6 cups	**orange juice**	1.35 l
1/4 cup	**light brown sugar**	60 mL
1/4 tsp	**cinnamon**	1 mL
2 tbsp	**cornstarch**	30 mL
2 tbsp	**water**	30 mL
3/4 tsp	**grated orange peel**	1 mL
	buttermilk or plain yogurt	

In a saucepan combine blueberries, orange juice, brown sugar and cinnamon. Mix cornstarch with water; gradually stir into blueberry mixture and cook until slightly thickened. Add orange peel. Pour into a bowl; refrigerate overnight. Serve cold with buttermilk swirled in or top with dollop of yogurt. Garnish with fresh blueberries and mint.

Blueberry Orange Sauce

YIELD 2 cups		
3 tbsp	**sugar**	45 mL
1 tbsp	**cornstarch**	15 mL
1/8 tsp	**salt** optional	0.5 mL
1/4 cup	**orange juice**	60 mL
1/4 cup	**water**	60 mL
1 cup	**blueberries** fresh or frozen	250 mL
1 cup	**orange sections**	250 mL

In a cup combine sugar, cornstarch and salt; set aside. In a small saucepan bring orange juice and water to a boil. Add blueberries and orange sections. Return to a boil; cook until liquid is released from fruit, about 2 minutes. Stir in sugar mixture; cook, stirring constantly, until sauce thickens, 1 to 2 minutes.

Carol Ann Shipman **Berries Cookbook**

Chilled Czech Blueberry Soup

Boil 2 cups (500 mL) of blueberries in the water. Stir in salt, cinnamon and sugar. Remove from heat. Whip flour into the sour cream, then whip both into the hot liquid. When well blended, return the pot to the heat and bring to a low simmer. Increase heat and stir until thickened. Remove from heat; stir in another 1/2 cup (125 mL) of blueberries, and chill in refrigerator. When ready to serve, stir in the remaining 1/2 cup (150 mL) of blueberries and ladle into bowls.

SERVES 4 – 6 (as first course)

3 cups	**blueberries** divided	750 mL
4 cups	**water**	1000 mL
	pinch of salt	
1 tbsp	**sugar**	15 mL
1-1/2 cups	**sour cream**	350 mL
3 tbsp	**flour**	45 mL
1/8 tsp	**cinnamon**	0.5 mL

Curried Chicken with BC Blueberries

Heat 1 tablespoon (15 mL) oil in large heavy skillet. Add chicken and sauté until golden. Remove chicken and set aside. Heat remaining tablespoon of oil in the pan and add the onions and grated ginger. Sauté until softened and fragrant. Add the coconut milk, chicken broth, curry paste and peppers; bring to a boil. Add the chicken, cook through. Stir in the blueberries and curry leaves; and season to taste with salt and pepper. Serve over steamed rice.

Curry Paste
Remove seeds from cardamom pods, discard pods. In a small heavy skillet, dry roast all the seeds and cloves together over low heat, stirring until fragrant, being careful not to burn them. Cool mixture and grind in an electric spice grinder, or clean coffee grinder, to a powder. Pour into a small bowl and stir in the turmeric and water to form a paste.

SERVES 2

2 tbsp	**vegetable oil**	30 mL
2	**skinned chicken breasts** boneless, cut into strips	2
1 cup	**minced onion**	250 mL
1 tbsp	**grated ginger**	15 mL
2 cups	**coconut milk**	500 mL
1 cup	**chicken broth**	250 mL
3	**hot red Indian peppers** seeded and slivered	3
1 cup	**blueberries** fresh or frozen	250 mL
2 tsp	**cilantro**	10 mL
6	**green curry leaves**	6
	salt and pepper to taste	

CURRY PASTE

2 tbsp	**coriander seeds**	30 mL
1 tsp	**cumin seeds**	5 mL
1 tsp	**fennel seeds**	5 mL
1 tsp	**mustard seed**	5 mL
6	**cardamom pods**	6
4	**cloves**	4
2 tsp	**turmeric**	10 mL
3 tbsp	**water**	45 mL

BLUEBERRIES

Blueberry-Onion Sauced Pork Tenderloin

SERVES 4

3/4 to 1 lb	**pork tenderloin**	455 g
2 tbsp	**butter**	30 mL
2	**medium onions** sliced	2
1/2 tsp	**salt**	2 mL
1/4 tsp	**ground black pepper**	1 mL
2 tbsp	**sugar**	30 mL
1/4 cup	**port wine** or **sweet sherry**	60 mL
2 tbsp	**balsamic vinegar**	30 mL
1 cup	**blueberries** fresh or frozen	250 mL
1 cup	**cherry tomatoes** chopped	250 mL

Preheat broiler. Broil pork, turning occasionally, until cooked through, about 20 minutes. Remove to a platter; cover to keep warm.

Meanwhile, in a large skillet over medium-high heat, melt the butter. Add onions, salt and pepper; cook until onions are golden, about 10 minutes. Add sugar; cook until onions are caramelized, 3 minutes longer. Add port, balsamic vinegar, blueberries and tomatoes; bring to a boil. Remove from heat. Thinly slice pork and serve with sauce.

Blueberry-Onion Sauced Pork Tenderloin

Veal Medallions with Blueberry-Citrus Sauce

In a bowl combine blueberries, orange juice, lemon juice, vermouth, orange peel, lemon peel and ginger; set aside. Melt butter in a large, heavy skillet and sauté veal until brown and just cooked through. Transfer veal to platter and keep warm. Add blueberry mixture to skillet and cook until mixture thickens, scraping up any brown bits (about 2 minutes). Spoon blueberry mixture over veal and serve.

SERVES 4

1 cup	**blueberries** fresh or frozen	250 mL
6 tbsp	**orange juice**	90 mL
6 tbsp	**lemon juice**	90 mL
2 tbsp	**vermouth**	30 mL
2 tsp	**orange peel** grated	10 mL
2 tsp	**lemon peel** grated	10 mL
1 tsp	**fresh ginger** minced	5 mL
3 tsp	**butter**	15 mL
1 lb	**veal medallions**	454 g
	salt and pepper to taste	

Veal Medallions with Blueberry-Citrus Sauce

Carol Ann Shipman **Berries Cookbook**

Hot and Sour Prawns with Blueberries

SERVES 4

1 lb	**prawn tails** peeled, butterflied	454 g
1/4 cup	**white wine**	60 mL
1 tbsp	**fresh ginger** grated	15 mL
1/2 cup	**chicken broth**	125 mL
2 tbsp	**soy sauce**	30 mL
2 tbsp	**tomato ketchup**	30 mL
1 tbsp	**cornstarch**	15 mL
1 tbsp	**rice vinegar**	15 mL
1 tbsp	**sugar**	15 mL
1 tsp	**sesame oil**	5 mL
1/4 tsp	**cayenne pepper**	1 mL
2 tbsp	**peanut oil**	30 mL
1	**bunch spinach** julienned	1
1	**red bell pepper** cut in 1-inch squares	1
1	**yellow pepper** cut in 1-inch squares	1
2	**cloves garlic** minced	2
6	**green onions** sliced diagonally	6
1 cup	**blueberries** fresh or frozen	250 mL

Combine prawns, 2 tablespoons (30 mL) white wine and ginger in a bowl. Cover and refrigerate 30 minutes.

Mix remaining wine, chicken stock, soy sauce, ketchup, cornstarch, vinegar, sugar, sesame oil and cayenne in a small bowl. Heat 2 teaspoons (10 mL) peanut oil in a wok or skillet over high heat. Add spinach to oil and stir-fry until spinach wilts. Transfer spinach to serving platter and keep warm.

Add 2 teaspoons (10 mL) oil, peppers and garlic to wok and stir-fry 1 minute. Add remaining oil, prawn mixture and green onions and stir-fry 2 minutes.

Stir in chicken broth mixture and blueberries and cook until sauce is clear, stirring frequently, about 2 minutes.

Variations
Substitute any firm white fish such as cod or skinless chicken breasts for the prawns.

Hot and Sour Prawns with Blueberries

Lake House Inn Blueberry Lasagna with Hazelnut Cream Sauce

In a large pot of boiling, salted water cook lasagna noodles until they are done. Drain and rinse under cold water. Return the noodles to the pot and add enough cold water to cover the noodles.

Preheat oven to 350°F (175°C) Wash and drain berries. In a large bowl combine the ricotta cheese, eggs and vanilla. Stir in the blueberries. Set aside.

Coat the bottom and sides of two 9 x 13 inch (22 x 33 cm) lasagna pans with cooking oil spray. Layer 4 noodles across the bottom of each pan overlapping slightly. Spread half of the filling over the noodles. Make another layer of noodles and filling. Sprinkle evenly with Crunchy Oat Topping.

Bake 45 minutes or until topping is golden brown and filling is bubbly. Remove from the oven and let lasagna set 15 to 20 minutes in the pan before cutting into squares.

Place lasagna square on a plate and spoon over some Hazelnut Cream Sauce.

Crunchy Oat Topping
In a medium bowl, use your hands to mix together all ingredients until coarse crumbs form. Spread evenly over the berry filling in each pan before baking.

Hazelnut Cream Sauce
Mix sour cream and hazelnut coffee creamer together until completely blended. Spoon over lasagna when ready.

SERVES 8 – 10

LASAGNA FILLING

16	lasagna noodles	16
4 cups	ricotta cheese	1000 mL
4	eggs	4
4 cups	fresh blueberries	1000 mL
2 tsp	vanilla extract	10 mL

CRUNCHY OAT TOPPING

6 tbsp	unsalted butter	90 mL
3/4 cup	all-purpose flour	175 mL
1 tsp	cinnamon	5 mL
1/2 cup	rolled quick oats	125 mL
3/4 cup	brown sugar firmLy packed	175 mL
1/2 cup	finely chopped nuts of your choice	125 mL

HAZELNUT CREAM SAUCE

1 cup	sour cream	250 mL
1 cup	hazelnut coffee creamer	250 mL

Spiced Blueberries

Spiced Blueberries

Combine blueberries, vinegar, lemon rind and sugar in a large non-reactive pot. Tie cloves and cinnamon in a cheesecloth bag and add to the blueberry mixture.

Bring to a boil, reduce heat and simmer 25 minutes uncovered.

Remove spice bag. Skim and discard any foam. Ladle berries into four clean and sterilized hot pint jars. Fill jars to 1/8 inch of the top.

Wipe jar rim and put lids on. Process in boiling water bath for 15 minutes. Count time from when the water returns to a boil.

Remove jars. Store for at least 3 weeks before using.

YIELD 8 cups

12 cups	**blueberries** fresh	3 kg
1 cup	**cider vinegar**	250 mL
1 cup	**sugar**	250 mL
1 tbsp	**grated lemon peel**	15 mL
2 tbsp	**whole cloves**	30 mL
1 inch	**piece cinnamon stick**	3 cm

Orange Buttered Rum Sauce on Blueberries

In large skillet combine orange juice, rum, sugar and grated peel. Bring to a boil over high heat. Cook, stirring occasionally, until liquid is reduced by half. Add lemon juice and butter; heat, stirring constantly, just until butter melts and is blended. Divide the berries among four serving dishes or long stemmed glasses. Pour warm sauce over. Serve immediately.

SERVES 4

1/2 cup	**orange juice**	125 mL
1/2 cup	**rum**	125 mL
3 tbsp	**light brown sugar**	45 mL
1 tsp	**grated orange peel**	5 mL
2 tsp	**lemon juice**	10 mL
4 tbsp	**cold unsalted butter** cut into 4 pieces	60 mL
1-1/2 cups	**blueberries**	375 mL
1 cup	**strawberries** cut into quarters	250 mL

Carol Ann Shipman **Berries Cookbook**

Blueberry Balsamic Vinegar

YIELD 5-1/2 cups

4 cups	**blueberries**	1000 mL
	frozen, thawed or fresh	
4 cups	**balsamic vinegar**	1000 mL
1/4 cup	**sugar**	60 mL
1	**lime peel**	1
	cut in strips from 1 lime (green part only)	
1	**cinnamon stick**	1
	3-inch piece (9 cm)	

In a large non-reactive saucepan crush blueberries with a potato masher or back of a heavy spoon. Add balsamic vinegar, sugar, lime peel and cinnamon; bring to a boil. Reduce heat and simmer covered for 20 minutes. Cool slightly and pour into a large bowl. Cover and refrigerate for 2 days.

Place a wire mesh strainer over a large bowl. In batches, ladle blueberry mixture into strainer, pressing out as much liquid as possible. Discard solids.

Pour vinegar into clean glass bottles or jars. Refrigerate, tightly covered, indefinitely. Use in salad dressings or drizzled over grilled chicken or beef.

Blueberry Vinaigrette

YIELD 1/2 cup

1/4 cup	**olive oil**	60 mL
3 tbsp	**blueberry balsamic vinegar**	45 mL
1/2 tsp	**salt**	2 mL
1/8 tsp	**ground black pepper**	0.5 mL

In a cup combine olive oil, blueberry vinegar, salt and pepper.

Variation

For a creamier dressing stir in 1 tablespoon (15 mL) mayonnaise or plain yogurt.

Blueberries with Creamy Banana Sauce

SERVES 4

1-1/2 cups	**blueberries**	375 mL
1/3 cup	**ricotta cheese**	75 mL
	skim or part-skim	
1 tsp	**lemon juice**	5 mL
1/3	**medium banana**	1/3
	sliced	
2 tbsp	**plain yogurt**	30 mL
1-1/2 tbsp	**sugar**	25 mL

In food processor or blender combine ricotta cheese, banana and all remaining ingredients except blueberries. Process or blend until very smooth, about 2 minutes.

Divide blueberries among 4 serving bowls or dishes. Spoon Creamy Banana Sauce over the berries. Sauce is best used within a day.

Blueberry Balsamic Vinegar

Maple-Ginger Frozen Cream with Blueberries

SERVES 4

FROZEN CREAM

1 cup	whipping cream	250 mL
1/4 cup	pure maple syrup, chilled	60 mL
3/4 tsp	ground ginger	4 mL
1	egg white, room temperature	1
1/4 cup	sugar	60 mL

TOPPING

1-1/4 cups	blueberries, divided	310 mL
2 tbsp	sugar	30 mL
1 tbsp	lemon juice	15 mL

Whip cream until soft peaks form. Gradually add the maple syrup, still whipping. Sprinkle the ginger over the cream and incorporate well. Refrigerate.

Whip the egg white to soft peaks. Continue whipping, gradually adding the sugar until stiff peaks form. Gently stir 1/3 of the whipped cream into the stiff egg whites; fold in the remaining whipped cream.

Divide the mixture among four mugs, cups or small serving bowls. Freeze 2 to 3 hours or overnight.

Blueberry Topping
Place 1 cup (250 mL) of the blueberries into a small saucepan with the sugar and lemon juice. Stir gently to combine.

Place the saucepan over medium-low heat and cover the pan. Cook gently for 2 to 3 minutes until the sugar has dissolved and the berries are tender but still whole. Remove from heat and refrigerate until ready to serve.

Serve
Spoon chilled blueberry topping on top of each dessert and garnish with the remaining raw blueberries.

Royal Blue Poached Pears with Blueberry Sauce

Pears
Combine all the ingredients in a large stockpot or saucepan. Add water just to cover the pears. Bring the liquid to a quick boil and immediately reduce heat to simmer. Simmer the pears for 45 minutes then test pears; hold a pear in a large spoon and stick a thin knife into the thickest part of the pear. Lift it up a little; the pear should slide slowly off the knife. When the pears are ready, remove from the heat and allow them to cool. Refrigerate in the liquid. They will hold for about 3 to 4 days.

Blueberry Sauce
Combine all the ingredients for the blueberry sauce in a medium-sized heavy bottom pan. Bring to a quick boil and turn down to a gentle simmer. Simmer for 10 to 12 minutes and remove from heat. Allow to cool. Mash blueberries through a fine strainer or cheesecloth. Refrigerate or freeze.

To Assemble
Remove a pear from the syrup and dry lightly with paper towel. Use a medium white dessert plate, drizzle blueberry sauce over the plate. Place poached pear in the middle of the blueberry sauce. Garnish with whipped cream rosettes.

SERVES 6

PEARS

6	whole Bartlett pears peeled, core removed, stems intact	6
1-1/2	sticks of cinnamon	1-1/2
5 - 6	whole cloves	5 - 6
1 cup	sugar	250 mL
1 cup	white wine	250 mL
1/2 tsp	turmeric or drops of yellow food-coloring	2 mL
	water to cover	

BLUEBERRY SAUCE

4 cups	blueberries fresh	1000 mL
1/2 cup	water	125 mL
1/4 cup	sugar or less depending on the sweetness of the berries	60 mL
1/2 tbsp	lemon juice	7 mL

Carol Ann Shipman **Berries Cookbook**

BLUEBERRIES

Berry Blue Frozen Dessert

SERVES 8 (1/2-cup servings)		
4 cups	**blueberries** fresh or frozen, divided	1000 mL
1/4 cup	**sugar**	60 mL
1 cup	**vanilla yogurt** 1% or skim (nonfat)	250 mL
1 tbsp	**lemon juice**	15 mL

Toss 2 cups (500 mL) blueberries with sugar. Let stand for 45 minutes, stirring occasionally.

Transfer berry-sugar mixture to food processor. Add yogurt and process until smooth. Strain through fine sieve.

Pour into 8 x 8 inch (20 x 20 cm) baking pan (or transfer to ice cream maker and process according to manufacturer's directions). Freeze uncovered until edges are solid.

Transfer to processor and blend until smooth again. Return to pan and freeze until edges are solid. Transfer to processor and blend until smooth again. Fold in remaining blueberries. Pour into plastic mold and freeze overnight. Let soften slightly to serve.

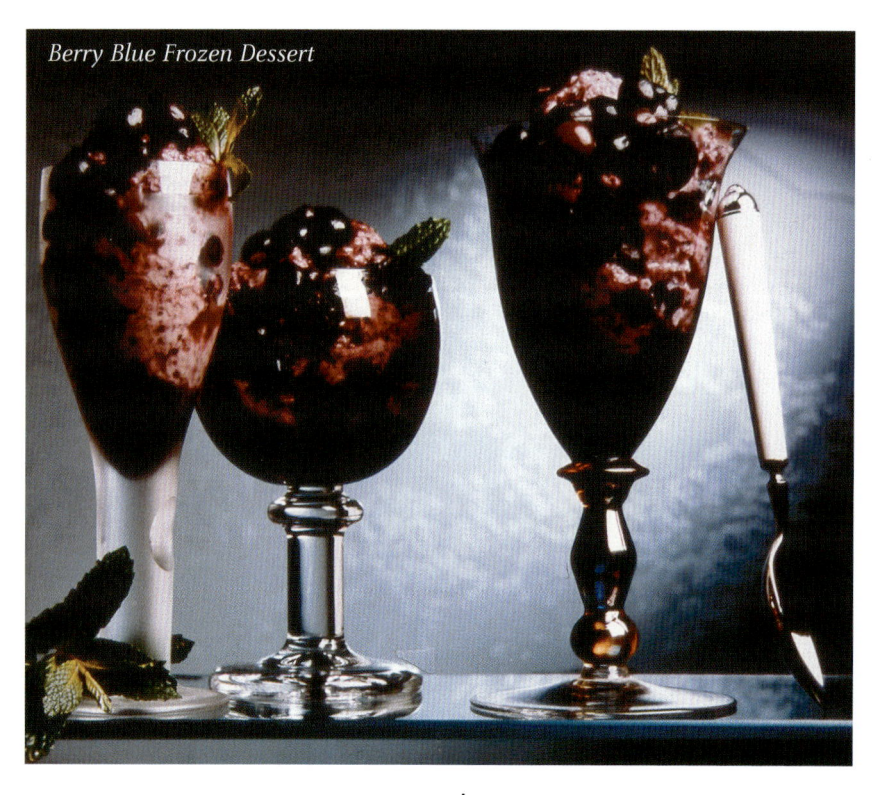

Berry Blue Frozen Dessert

Fresh Blueberry & Lemon Parfait

Prepare instant lemon pudding according to package directions, using the milk. In a medium bowl, with an electric mixer at medium-high speed, beat whipping cream until soft peaks form. Fold whipped cream into prepared lemon pudding. In either 4 to 6 individual serving glasses or a bowl, spoon a layer of the pudding mixture; sprinkle lightly with cookies and a layer of blueberries. Repeat layers one more time, ending with the pudding. Refrigerate, covered, for about 30 minutes. Garnish with mint sprigs and blueberries, if desired.

SERVES 4 – 6

1	**medium package** instant lemon pudding	106 g
1-1/2 cups	**milk**	375 mL
1 cup	**whipping cream**	250 mL
1 cup	**gingersnap cookies** coarsely crushed	250 mL
2-1/2 cups	**blueberries** fresh	625 mL

Orange Almond Blueberries

In a small saucepan combine orange juice, sugar and almond extract. Heat to boiling, then reduce heat and simmer until syrup is reduced by half. Divide the berries among four serving dishes. Pour orange syrup over the berries and sprinkle with almonds. Serve immediately.

This sauce also works well over other berries, or try serving the berries and sauce over angel food or pound cake.

SERVES 4

1 cup	**orange juice** fresh	250 mL
1/4 cup	**light brown sugar** packed	60 mL
1 tsp	**almond extract**	5 mL
1-1/2 cups.	**blueberries**	375 mL
1 tbsp	**sliced almonds** toasted	15 mL

Carol Ann Shipman **Berries Cookbook**

BLUEBERRIES

Blueberry Mincemeat

Combine all ingredients except blueberries in a large mixing bowl.

Stir in blueberries. Pack closely in sterilized jars and cover tightly; refrigerate until used (1 week). For longer storage, freeze.

YIELD 6 cups

1-1/4 cups	sultana raisins	300 mL
1-1/4 cups	golden raisins	300 mL
1/2 cup	dried cranberries	125 mL
1/4 cup	mixed peel	60 mL
1/2 cup	brown sugar	125 mL
1/2 cup	margarine	125 mL
1 tsp	grated lemon peel (yellow part only)	5 mL
2 tbsp	brandy	30 mL
1 tbsp	lemon juice	15 mL
1 tsp	ground cinnamon	5 mL
1/2 tsp	ground cloves	2 mL
1/2 tsp	ground ginger	2 mL
1/2 tsp	ground nutmeg	2 mL
4 cups	blueberries fresh or frozen	1000 mL

Blueberry Mincemeat

Blueberry & Roasted Corn Salsa

Soak unshucked corn in a large bowl of water at least one hour. Heat barbecue. Roast corn until husks are charred, turning frequently for about 10 minutes. Shuck the corn and cut from the cob. Combine all the ingredients in a large bowl. Season with salt and pepper to taste. Add blueberries.

YIELD 4 cups

2	ears of corn with husks (do not shuck)	2
1/2	medium red pepper	1/2
1	large jalapeño pepper finely chopped	1
1/2	red onion minced	1/2
5 tbsp	lime juice	75 mL
2-1/2 tbsp	olive oil	40 mL
2	cloves garlic minced	2
1/4 cup	cilantro chopped	60 mL
1 cup	blueberries	250 mL

BLUEBERRIES

Blueberry Wild Rice Salad

Wash the wild rice. Combine the rice with chicken broth and enough water to cover by 1 inch in a medium saucepan. Bring to a boil. Reduce heat; cover and simmer until rice is tender, about 40 minutes. Drain and cool. Stir in hazelnuts, cranberries, apricots and red onion.

Dressing
In a small bowl, whisk together the lime juice, honey, grated ginger, lime and lemon peel. Gradually whisk in the olive oil. Season to taste with salt and pepper.

Pour over the rice and mix well. Gently fold in the blueberries. Arrange greens on a platter and mound rice on top.

SERVES 4

1 cup	**wild rice**	250 mL
2 cups	**chicken broth**	500 mL
1/4 cup	**hazelnuts** chopped, toasted	60 mL
1/2 cup	**cranberries** dried	125 mL
3/4 cup	**apricots** chopped dried	175 mL
1/2 cup	**red onion** chopped	125 mL
1/4 cup	**lime juice**	60 mL
2 tbsp.	**honey**	30 mL
1-1/2 tsp.	**ginger** grated fresh	7 mL
1 tsp.	**grated lime peel**	5 mL
6 tbsp.	**olive oil**	90 mL
	salt & pepper	
1 cup	**blueberries**	250 mL
8 oz.	**mixed greens**	250 mL

Blueberry Tortilla Pizza

Preheat broiler. In a small bowl, combine ricotta cheese and sugar and set aside. In a small bowl, combine blueberries and strawberries. Arrange tortilla on a broiler pan; brush with butter and sprinkle with cinnamon sugar. Broil about 6 inches from heat source, until lightly browned, about 2 minutes. Cool slightly. Spread ricotta mixture over tortilla, top with blueberry mixture and then sprinkle with coconut.

Toasted Coconut
To toast coconut, place in a skillet over moderate heat until pale gold, stirring constantly.

SERVES 4

1/2 cup	**ricotta cheese** or **whipped cream cheese**	125 mL
1 tbsp	**sugar**	15 mL
2 cups	**blueberries** fresh	500 mL
1/2 cup	**strawberries** sliced	125 mL
1	**flour tortilla** 10-inch	1
1 tbsp	**butter** melted	15 mL
2 tsp	**equal amounts of sugar and cinnamon**	10 mL
1/4 cup	**coconut** toasted, shredded	60 mL

Carol Ann Shipman **Berries Cookbook**

strawberries

Look for a brilliant, even, red color and symmetrical shape. While you should always rinse off fruit before eating, it's also wise to look for strawberries that are clean in the first place. To keep strawberries from absorbing large quantities of water, hull after washing. For best flavor allow strawberries to reach room temperature before serving. They should look fresh and unwilted. At home keep your strawberries refrigerated in a ventilated container.

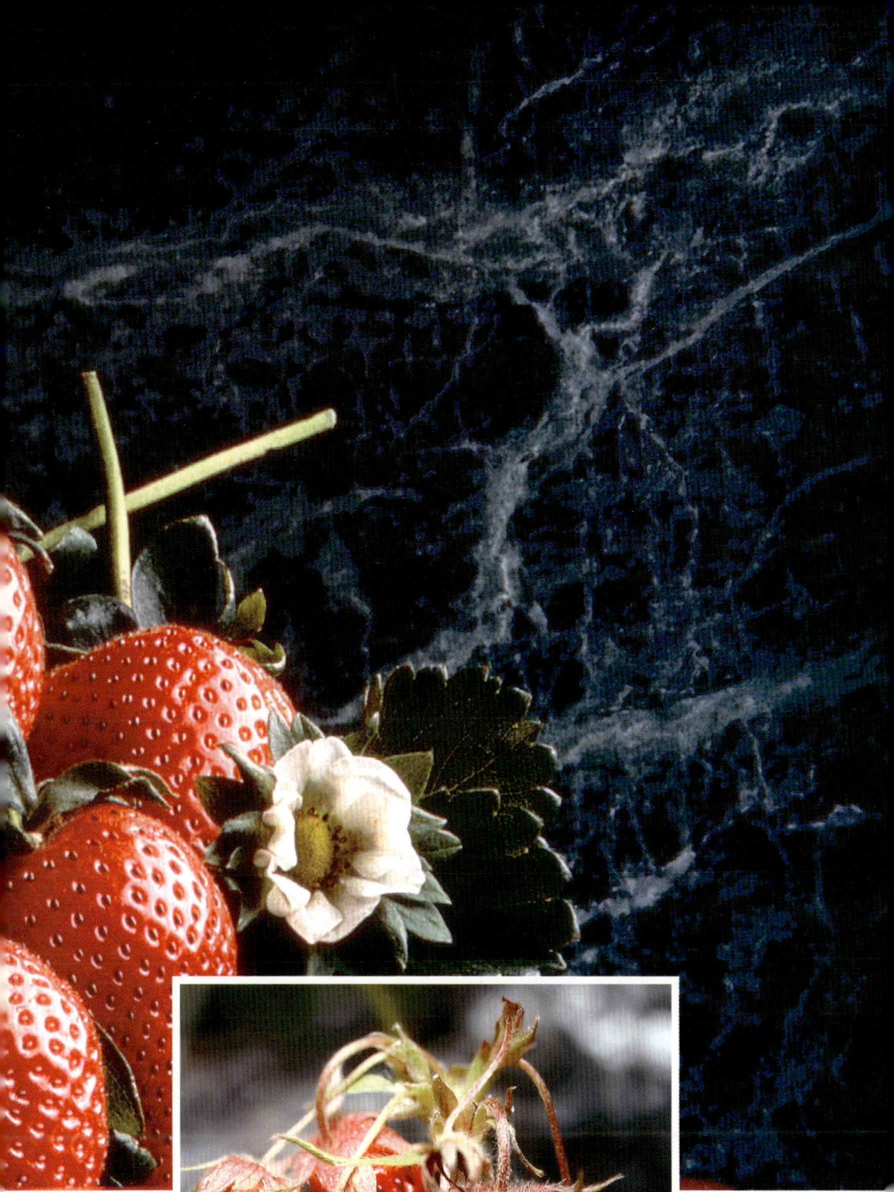

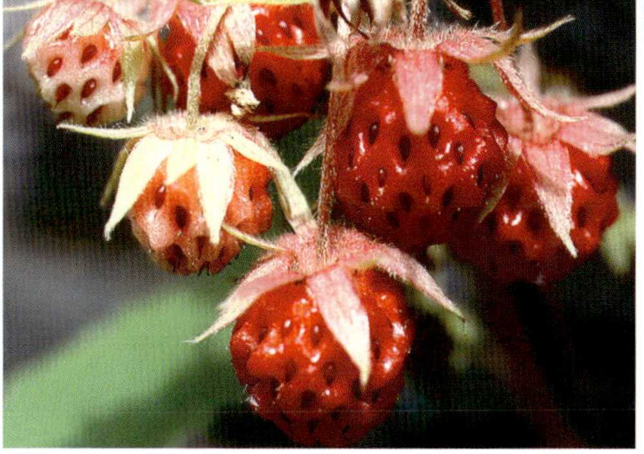

STRAWBERRIES

Strawberry Sangria Ice

SERVES 4

4 cups	**strawberries** stemmed and divided	1000 mL
1/4 cup	**sugar**	60 mL
1 cup	**dry red wine**	250 mL
2 tbsp	**lemon juice**	30 mL
1-1/2 tbsp	**orange juice** frozen concentrate	22 mL
	sparking water	
	orange slices	
	mint sprigs for garnish	

In large bowl crush 1/2 of the strawberries with the sugar. Add wine, lemon juice and thawed orange juice concentrate, stirring to dissolve sugar. Pour into shallow pan; freeze until firm.

Presentation

Fill 4 stemmed glasses with spoonfuls of sangria ice and the remaining strawberries, halved (reserve 4 whole strawberries for garnish). Fill with sparking water. Garnish with whole strawberries, orange slices and mint sprigs.

Serve with spoons and straws.

Strawberry Cream

SERVES 1

1 oz	**light rum**	30 mL
2 oz	**Strawberry Schnapps**	60 mL
2 oz	**half and half cream**	60 mL
1/2 tsp	**superfine sugar**	2 mL
3	**strawberries**	3

Combine all ingredients in a blender, blend on high speed for 1 minute. Pour into large red wine glasses.

Strawberry Malted Milk Shake

SERVES 2

2 cups	**strawberries** halved	500 mL
2-1/2 cups	**vanilla ice cream**	625 mL
1/3 cup	**milk**	75 mL
3-1/2 tbsp	**malted milk powder**	52 mL
1 tbsp	**sugar**	15 mL
	whole strawberries with leaves	

Combine all ingredients in a blender, process until milkshake consistency. Garnish with strawberry and long straw.

Strawberry Sangria Ice

STRAWBERRIES

Strawberry Smoothies

Blend ingredients in processor until smooth. Pour into glasses.

SERVES 2 large or 4 small

2 cups	**strawberries** fresh frozen	500 mL
2	**frozen bananas** keep frozen bananas on hand	2
1/2 cup	**orange juice**	125 mL
1/2 cup	**yogurt**	125 mL

Strawberry Wine Punch

In a punch bowl, combine strawberries, sugar and 2 cups (500 mL) of rosé wine. Cover and let stand at room temperature for 1 hour. Before serving punch, add frozen lemonade concentrate and pineapple juice. Stir until lemonade is thawed. Stir in remaining wine and club soda. Add ice or ice ring.

SERVES 12

3-3/4 cups	**strawberries** fresh frozen, sliced	925 mL
1/2 cup	**sugar**	125 mL
1 bottle	**rosé wine**	750 mL
6 oz	**lemonade** frozen concentrate	180 mL
1 cup	**pineapple juice** chilled	250 mL
28 oz	**club soda** chilled	840 mL
	ice or ice ring	

Strawberry Colada

Combine all ingredients in a blender on high speed. Pour into tall glasses, garnish with pineapple wedges.

SERVES 1

1 oz	**light rum**	30 mL
1 oz	**dark rum**	30 mL
1 oz	**coconut cream**	30 mL
1 cup	**ice**	250 mL
1/4 cup	**pineapple juice**	125 mL
4	**strawberries**	4

Strawberry Banana Cocktail

Combine all ingredients in a blender on high speed for 1 minute. Pour into a parfait glass.

SERVES 1

2 oz	**light rum**	60 mL
3 oz	**skim milk**	90 mL
1 oz	**apple juice**	30 mL
1	**banana**	1
4	**strawberries**	4

Savory Strawberries on Lime Ice

Lime Ice
Combine lemon-lime soda and grated lime peel. Pour into 8-inch (20 x 20 cm) square pan and freeze 6 hours or overnight. Remove from pan and process in food processor or blender until smooth. Pour into pan and re-freeze at least 4 hours.

Savory Strawberries
Combine wine, basil and lime peel; mix well. Pour over strawberries and marinate 1 to 2 hours. Spoon over Lime Ice.

SERVES 4

SAVORY STRAWBERRIES

1/2 cup	**fruity white wine**	125 mL
1/4 cup	**fresh basil** chopped	60 mL
1 tsp	**lime peel** grated	5 mL
4 cups	**strawberries** fresh, halved	1000 mL

LIME ICE

4 cups	**lemon-lime soda**	1000 mL
1 tsp	**lime peel** grated	5 mL

Strawberry Fool

Process the berries in food processor with the sugar, pulsing until just mixed. Add orange extract. Whip the cream until stiff, then fold in the strawberry mixture. Transfer to serving bowl and refrigerate at least 4 hours, or until firm. This can also be placed into individual parfait glasses and served with sugar cookies for a fancier presentation.

SERVES 6

1-1/2 cups	**strawberries** fresh frozen, chopped	375 mL
1/4 cup	**sugar**	60 mL
1/2 tsp	**orange extract**	2 mL
1-1/4 cups	**whipping cream**	300 mL

Strawberry Daiquiri

Place lemonade, water, rum and sugar in blender and mix well. Add strawberries and ice cream and mix slightly.

SERVES 6

12 oz	**pink lemonade** frozen	360 mL
7 oz	**water or ice cubes**	210 mL
7 oz	**rum**	210 mL
1 tbsp	**sugar**	15 mL
7 oz	**strawberries** fresh frozen	210 mL
2 cups	**strawberry ice cream**	500 mL

Carol Ann Shipman **Berries Cookbook**

Summertime Strawberry Antipasto

Summertime Strawberry Antipasto

To make vinaigrette, whisk together oil, water, vinegar, mint, mustard, honey, garlic, salt and pepper. To serve antipasto, arrange remaining ingredients, dividing equally among individual serving plates (or arrange on platter).

Spoon vinaigrette over antipasto.

Serve immediately.

SERVES 6 appetizers

3 tbsp	**olive oil**	45 mL
3 tbsp	**water**	45 mL
3 tbsp	**red wine vinegar**	45 mL
2 tbsp	**fresh mint** chopped or 2 tsp dried mint, crumbled	30 mL
2 tsp	**dijon-style mustard**	10 mL
2 tsp	**honey**	10 mL
1	**clove garlic** minced	
1/4 tsp	**each salt and pepper**	1 mL
2 cups	**strawberries**	500 mL
1/2	**small cantaloupe** seeded and cut into 1/2-inch thin wedges	1/2
1 cup	**red and/or green seedless grapes**	250 mL
2	**nectarines or plums** pitted and cut into wedges	2
1/4 lb	**asiago or parmesan cheese** cut into thin slices	114 g
1/4 lb	**prosciutto or ham** very thinly sliced	114 g

Strawberry Cheese Tarts

Combine cream cheese, sour cream, sugar and lemon peel in small bowl; beat until smooth. Spread evenly in crumb crusts. Refrigerate about 4 hours. Before serving, spread the strawberry purée evenly over the filling.

SERVES 12

1 lb	**cream cheese** low-fat, softened	454 g
1/2 cup	**sour cream** low-fat	125 mL
3 tbsp	**sugar**	45 mL
1 tsp	**lemon peel** grated	5 mL
	package of 12 (3-inch) graham cracker tart shells	
2/3 cup	**strawberries** fresh, crushed in food processor	150 mL

STRAWBERRIES

Calypso Strawberry-Mango Salsa

YIELD 3 cups

1-1/2 cup	**strawberries** chopped	350 mL
1	**large mango**	1
1/4 cup	**green onions** sliced, with tops	60 mL
2 tbsp	**fresh lime juice**	30 mL
1 tbsp	**chopped cilantro**	15 mL
1/2 tsp	**red pepper flakes**	2 mL
1/4 tsp	**ground cumin**	1 mL
	salt to taste	

In bowl, toss all ingredients except salt to blend thoroughly. Mix in salt.

Serve immediately or cover and refrigerate up to two days. Serve with corn chips, chicken or mild white fish.

Strawberry Nachos

SERVES 6

3 cups	**strawberries** sliced	750 mL
1/3 cup	**sugar**	75 mL
1/4 cup	**Amaretto** almond flavored liqueur	60 mL
1/2 cup	**sour cream** nonfat	125 mL
1/2 cup	**frozen** **reduced-calorie whipped** **topping, thawed**	125 mL
2 tbsp	**sugar**	30 mL
6	**(7-inch) flour tortillas** cut into 8 wedges	6
	butter-flavored **vegetable cooking spray**	
2 tsp	**cinnamon-sugar**	10 mL
2 tbsp	**sliced almonds** toasted	30 mL
2 tsp	**chocolate** shaved, semi-sweet	10 mL

Combine strawberries, 1/3 cup (75 mL) sugar and Amaretto in a bowl; stir well. Cover and chill 30 minutes. Drain, reserving juice for another use.

Combine sour cream, whipped topping, 2 tablespoons (30 mL) sugar, and cinnamon in a bowl; stir well. Cover and chill.

Arrange tortilla wedges on 2 baking sheets; lightly coat with cooking spray. Sprinkle evenly with cinnamon-sugar. Bake at 400°F (200°C) for 7 minutes or until crisp. Cool on wire rack.

To Serve
Arrange 8 tortilla wedges on a serving plate; top with about 1/3 cup (75 mL) strawberry mixture and 2-1/2 tablespoons (32 mL) sour cream mixture. Sprinkle with almonds and chocolate.

Strawberry Club Sandwiches

In small bowl beat cheese, orange juice and honey. Blend thoroughly; mix in walnuts, set aside. Cut cake into 12 equal slices. Slice some of the strawberries in half for plate decoration and slice the remainder for sandwiches. Lay out 12 slices of cake, each sandwich will take three pieces of cake. Place small amount of cream cheese mixture, then a layer of strawberry slices on two pieces of cake. Sandwich them together with the third piece as you would a clubhouse sandwich. Cut diagonally into halves; skewer each with a sandwich pick. Serve sandwiches with remaining strawberries halved, dividing equally on each plate. Dollop with whipped cream; garnish with mint sprigs.

SERVES 4

6 oz.	**light cream cheese** softened	170 g
2 tbsp	**orange juice** frozen concentrate, thawed	30 mL
1 tbsp	**honey**	15 mL
1/3 cup	**toasted walnuts** chopped	75 mL
1	**pound cake loaf** preheated	12 oz
4 cups	**strawberries** fresh	1000 mL
1/2 cup	**whipping cream** whipped, lightly sweetened	125 mL

Strawberry Three Cheese Mold

Line a two-cup bowl with plastic wrap, letting the plastic overhang the edges somewhat, set aside. Put the three cheeses in a mixing bowl and combine. Add the vinegar and combine well. Add salt to taste. Give it a few grinds of fresh pepper. Transfer the mixture to the plastic-lined bowl, cover and refrigerate until firm. A half hour before serving remove the mold from the refrigerator to soften slightly. Turn out onto a serving platter. Leave the plastic wrap on.

To Serve
Remove plastic wrap. Unmold. Press strawberry slices all around perimeter and over surface in an attractive fashion. Serve with crackers.

SERVES 10–12 for a party cheeseboard

1 cup	**cream cheese** room temperature	250 mL
1/4 cup	**blue cheese** crumbled, room temperature	60 mL
1/2 cup	**goat cheese** room temperature (log type)	125 mL
1 tbsp	**balsamic vinegar**	15 mL
	salt and fresh ground pepper to taste	
2 cups	**strawberries** sliced	500 mL
2	**basil leaves** rinsed and dried, minced	2

Carol Ann Shipman **Berries Cookbook**

California Skinny Dips

STRAWBERRIES

California Skinny Dips

For each serving, wash and pat dry with paper towels about 3/4 cup (175 mL) strawberries; for each dip, whisk ingredients until smooth in separate containers.

To Serve
Make one or more of the dips. Serve in small bowls to accompany strawberries.

SERVES 6

| 2-1/4 cups | **strawberries** divided | 560 mL |

CHOCOLATE FUDGE DIP

6 tbsp	**nonfat yogurt**	90 mL
6 tbsp	**prepared chocolate fudge sauce**	90 mL
1-1/2 tsp	**orange juice** frozen concentrate, thawed	7 mL

HONEY ALMOND DIP

2/3 cup	**nonfat yogurt**	150 mL
3 tbsp	**almonds** toasted, slivered, finely chopped	45 mL
2-1/2 tbsp	**honey**	52 mL

STRAWBERRY CREAM DIP

1/2 cup	**sour cream** low-fat	125 mL
1/4 cup	**strawberries** no sugar added	60 mL
	fruit spread or strawberry jam	

Cinnamon Glazed Strawberries

In medium saucepan combine sugar, butter, lemon juice and cinnamon. Cook and stir over medium heat until syrupy and thick. Remove from heat and add berries, tossing gently to coat each berry. Serve immediately.

Variations
Add coconut along with berries, toss to coat. Cool glazed berries on a plate or cookie sheet, then roll in coconut, or chocolate or both.

SERVES 4

2 tbsp	**sugar**	30 mL
1 tbsp	**butter**	15 mL
2 tsp	**lemon juice**	10 mL
1/2 tsp	**ground cinnamon**	2 mL
2 cups	**strawberries** fresh	500 mL

VARIATIONS

| 2 tbsp | **toasted shredded coconut** | 30 mL |
| 2 tbsp | **chocolate sprinkles** or finely grated semi-sweet chocolate | 30 mL |

Taste of Summer Strawberry Chutney

SERVES 8

1/2 cup	**golden raisins**	125 mL
1/2 cup	**dark brown sugar** firmLy packed	125 mL
1/2 cup	**strawberry preserves**	125 mL
1/2 cup	**strawberry wine vinegar**	125 mL
1/2 cup	**orange juice** fresh	125 mL
2 tsp	**ginger root** peeled, minced	10 mL
1/2 tsp	**curry powder**	2 mL
1	**medium navel orange**	1
4 cups	**strawberries** diced	1000 mL
1/2 cup	**sliced almonds**	125 mL

Combine the first 8 ingredients in large heavy saucepan; bring to boil. Cook uncovered, over medium heat 15 minutes or until slightly thickened and syrupy, stirring frequently. Add strawberries; reduce heat, and simmer, uncovered 10 minutes or until thickened, stirring occasionally. Remove from heat; stir in almonds. Spoon into a bowl; cover and chill.

The Colony Hotel Famous Strawberry Soup

SERVES 6

2 cups	**fresh strawberries**	500 mL
1 cup	**frozen strawberries** and juice	250 mL
1/2 cup	**orange juice** freshly squeezed	125 mL
1/2 cup	**sugar**	125 mL
1 cup	**plain yogurt** nonfat	250 mL
1/2 cup	**sour cream** low-fat	125 mL
1/2 cup	**water**	125 mL
1/2 cup	**half & half cream**	125 mL
2/3 cup	**strawberry liqueur** (optional)	150 mL
6	**fresh strawberries** with leaves on for garnish	6

Purée the fresh and frozen strawberries in a blender or food processor until finely chopped. Pour the strawberries into a mixing bowl. Add the remaining ingredients (except garnish) and whisk thoroughly. Chill for 2 hours. Cut each strawberry for the garnish into several thin vertical slices that measure three-quarters of the way to the stem. Fan out the strawberry slices on top of each bowl of soup.

STRAWBERRIES

Chef Cushman's Strawberry Chicken Salad with Hoisin-Sesame Dressing

Rinse, drain, dry and hull strawberries and cut into halves or thick slices. In large bowl combine all salad ingredients with about 1/2 cup (125 mL) Hoisin-sesame dressing and toss lightly to coat greens and berries. Divide among 6 chilled salad plates to serve.

Hoisin-Sesame Dressing
Put all ingredients except oil in blender container or food processor and blend until shallots and ginger are very finely chopped. With motor running, gradually add oil until smooth. Store dressing in refrigerator in tightly covered container.

If not using blender or food processor, grate ginger and mince garlic and shallots, then whisk together with remaining ingredients.

SERVES 6

STRAWBERRY CHICKEN SALAD

2 cups	fresh strawberries	500 mL
12 oz	smoked chicken or cooked skinless chicken breasts, julienned	340 g
3/4 cup	jicama julienned	175 mL
3/4 cup	fennel bulb very thin sliced	175 mL
2	sliced green onions	2
8 cups	mixed baby greens washed, dried and chilled	2000 mL

HOISIN-SESAME DRESSING

1-2 inch piece	fresh ginger peeled	1-2 inch piece
1	peeled garlic clove	1
1	medium shallot peeled	1
1/2 cup	rice vinegar	125 mL
1/2 cup	light soy sauce	125 mL
1/4 cup	sugar	60 mL
2 tbsp	Hoisin sauce	30 mL
2 tbsp	sesame seeds toasted	30 mL
1 tbsp	sesame oil	15 mL
1 tbsp	dijon mustard	15 mL
1 cup	vegetable oil	250 mL
	dash ground black pepper	

Drunken Fruit Salad

Combine all of the fruit, coconut, and almonds in a large bowl.

In a small mixing bowl, stir the sugar into the rum and brandy until dissolved. Pour over fruit mixture and combine.

Chill at least 2 hours, stirring occasionally.

SERVES 6

2	bananas peeled and sliced	2
2 cups	cubed watermelon seeds removed	500 mL
2	oranges peeled and sectioned	2
1	grapefruit peeled and sectioned	1
12	strawberries thinly sliced	12
1/3 cup	fresh coconut	75 mL
1/3 cup	slivered almonds	75 mL
1/4 cup	dark rum	60 mL
1/4 cup	brandy	60 mL
2 tbsp	sugar	30 mL

Carol Ann Shipman *Berries Cookbook*

Strawberry Cheesecake

Light and Luscious Strawberry Cheesecake with Fresh Strawberry Sauce

Preheat oven to 300°F (150°C). In medium bowl mix crumbs and margarine. Press onto bottom and 2 inches up sides of lightly greased 9-inch (22 cm) springform pan; set aside.

Filling
In mixer bowl beat ricotta cheese until smooth. Add 3/4 (175 mL) cup of the sugar, the flour, egg yolks, lemon peel and vanilla; mix well. Stir in sour cream substitute to blend thoroughly. In another bowl, beat egg whites until stiff but not dry; fold into cheese mixture. Pour into prepared crust; smooth top. Bake 1 hour. Turn off oven; cool in oven 1 hour with door ajar. Remove from oven; chill thoroughly.

Sauce
In blender or food processor purée 1/2 of the strawberries with the remaining 1/4 cup (60 mL) sugar and the lemon juice; strain sauce to remove seeds. Cover and chill. To complete cake, halve remaining strawberries; arrange on top of cake. Brush strawberries with jelly. Cut cake into wedges; serve with sauce.

SERVES 12

1-1/2 cups	**graham cracker crumbs**	350 mL
3 tbsp	**melted margarine** or butter	45 mL
15 oz	**ricotta cheese** part-skim (1 carton)	475 g
1 cup	**sugar** divided	250 mL
2/3 cup	**flour**	150 mL
4	**eggs** separated	4
2 tbsp	**vanilla**	30 mL
1 cup	**nonfat sour cream substitute**	250 mL
6 cups	**strawberries**	1500 mL
4 tsp	**lemon juice**	20 mL
1/4 cup	**red currant jelly** melted	60 mL

Grilled Chicken Breasts with Fresh Strawberries in Strawberry Red Wine Balsamic Sauce

SERVES 4

2 cups	**strawberries** fresh, divided	500 mL
2 tsp	**light olive oil** or vegetable oil	10 mL
2/3 cup	**sliced shallots**	150 mL
1	**clove garlic** crushed	1
1	**small bay leaf**	1
1/2 tsp	**dried thyme leaves**	2 mL
2 tbsp	**balsamic vinegar**	30 mL
1/3 cup	**dry red wine** full-bodied	75 mL
3/4 cup	**chicken broth**	175 mL
1-1/2 tsp	**cornstarch** mixed with 2 tsp cold water	7 mL
	salt and pepper to taste	
4	**boneless chicken breasts** (4-5 ounces each)	4

Rinse, dry, hull and slice berries. In a small saucepan over medium-high heat, heat oil. When oil is hot, add shallots, garlic, bay leaf and thyme. Cook, stirring occasionally, until lightly browned, about 2 to 3 minutes. Add 3/4 cup (175 mL) of the sliced strawberries and cook for an additional minute. Add balsamic vinegar. Boil mixture for one minute. Add red wine, reduce heat to medium and simmer until liquid reaches to just below strawberries, about 5–8 minutes. Add broth and simmer 5 minutes. Stir. Bring liquid to a boil.

Whisk in cornstarch mixture and boil for 30 seconds. Remove from heat. Strain sauce into another small saucepan. Season with salt and pepper to taste. Set aside.

Prepare chicken breasts over medium-high heat on your grill. Place breasts on individual plates. Scatter remaining strawberries over and around breasts. Spoon sauce over each breast.

Strawberry Whipped Cream

SERVES 1

3-4	**large strawberries**	3-4
1/4 cup	**whipping cream**	60 mL
2 tsp	**confectioners' sugar**	10 mL

Mash berries with a fork. You should have 1/4 cup (60 mL) mashed berries. Set aside. Whip cream with sugar to soft peaks. Add mashed berries and continue whipping until stiff peaks. Serve immediately or refrigerate one hour.

STRAWBERRIES

Chef Revsin's Angel Hair Pasta with Strawberry and Brown Sugar Sauce

In small bowl, toss strawberries and 1 tablespoon (15 mL) brown sugar, cover and set aside.

Cook pasta according to directions, omitting salt. Drain well. Set aside.

In medium saucepan, combine cream and angel hair pasta. Bring to boil, stirring constantly. Reduce heat and simmer several minutes, add remainder of the brown sugar, simmer another 5 minutes until sugar is dissolved and cream is light brown in color and smooth in consistency. Remove from heat. Add cooked pasta to sauce until it is well coated and pasta is heated thoroughly. Place pasta in serving bowl or individual dishes. Spoon strawberries and juices over pasta. Sprinkle with chopped nuts

SERVES 4

1 cup	**strawberries** thinly sliced	250 mL
1/3 cup	**dark brown sugar** divided	75 mL
3/4 cup	**whipping cream**	175 mL
4 oz	**angel hair pasta**	125 g.
2 tbsp	**toasted hazelnuts** chopped	30 mL

Strawberry Ricotta Dip

Place all ingredients except strawberries into a food processor. Process well until very smooth. Add strawberries just to combine. Refrigerate until well chilled and thickened.

Variation

Substitute with blueberry jelly, also 1/4 cup (60 mL) blueberries. Add all other ingredients. (Follow recipe).

Variation

Substitute with raspberry jelly, also 1/4 cup (60 mL) raspberries. Add all other ingredients. (Follow recipe).

YIELD 1 cup

1/2 cup	**ricotta** part skim	125 mL
3 tbsp	**plain yogurt**	45mL
3 tbsp	**strawberry jelly**	45 mL
1/8 tsp	**vanilla extract**	0.5 mL
1/4 cup	**strawberries** lightly crushed with a fork	60 mL

Carol Ann Shipman **Berries Cookbook**

Strawberry-Almond Cream Napoleons

Strawberry-Almond Cream Napoleons

Cook puff pastry according to package directions. Set aside.

Heat oven to 400°F (200°C). Thaw folded pastry sheets and cut along folds to make three equal strips. Bake about 15 minutes until well browned and baked through. Cool.

Meanwhile in a bowl, whisk pudding mix and milk until thickened. Fold in whipped topping to blend thoroughly. Refrigerate. Carefully split each piece of pastry in half; cover bottom halves with almonds, then pudding mixture and strawberries, dividing equally. Cover with pastry tops. Sprinkle with confectioners' sugar. Serve with strawberry sauce.

All components of this dessert can be prepared in advance. Cover and refrigerate pudding mixture and strawberries; assemble just before serving.

Strawberry Sauce
In a blender purée strawberries with lemon juice and sugar until smooth.

SERVES 6

1 sheet	**puff pastry** (half of 17-ounce package)	1 sheet
1 pkg	**instant vanilla pudding mix** (3.4 ounces)	106 g
1 cup	**2% milk**	250 mL
1 tsp	**almond extract**	5 mL
1 cup	**light whipped topping**	250 mL
2 cups	**strawberries** sliced	500 mL
1/3 cup	**sliced almonds** toasted	75 mL
	confectioners' sugar for garnish	

STRAWBERRY SAUCE

1 cup	**strawberries**	250 mL
1 tsp	**lemon juice**	5 mL
1 tbsp	**sugar**	15 mL

Classic Strawberry Pie

Classic Strawberry Pie

In bowl, whisk pudding mix and milk 2 minutes; chill 30 minutes. Spread pudding evenly in piecrust. Cover top of pie with strawberries, pointed ends up. Dust lightly with powdered sugar.

Graham Cracker Crust
Heat oven to 375°F (190°C). In bowl mix graham cracker crumbs with the sugar and melted margarine or butter until thoroughly blended. Press crumb mixture firmly onto bottom and side of 9-inch (20 cm) pie plate. Bake in center of oven about 8 minutes until lightly brown; cool.

SERVES 6 – 8

1 pkg	**instant vanilla pudding mix** (3.4 ounces)	106 g
1 cup	**skim milk**	250 mL
1	**9-inch prepared graham cracker piecrust** or use crust recipe below	1
4 cups	**strawberries**	1000 mL
	confectioners' sugar for dusting	

GRAHAM CRUST

1-1/4 cups	**graham crumbs**	300 mL
1/4 cup	**sugar**	60 mL
1/3 cup	**butter or margarine** melted	75 mL

Strawberry-Hot Fudge Sundae Cake

In mixer bowl combine flour, baking powder, 2 tablespoons of the cocoa powder and 2/3 cup of the granulated sugar. Add milk, butter and vanilla; beat until smooth and well blended. Spread in greased 8-inch square baking pan; set aside. In small saucepan combine the remaining white sugar, adding the remaining 6 tablespoons cocoa powder, brown sugar and water. Bring to boil, stirring to dissolve sugar. Gently spoon over cake batter. Bake in 350°F (175°C) oven about 35 minutes until springy to the touch. While cake is baking, slice strawberries into a bowl and toss with additional sugar to sweeten. To serve, spoon warm cake and fudge sauce into individual bowls. Add scoops of ice cream and sweetened strawberries. Cake can be prepared ahead and served cooled or reheated in a low oven.

SERVES 6

1 cup	**flour**	250 mL
2 tsp	**baking powder**	10 mL
3/4 cup	**sugar**	175 mL
1/2 cup	**cocoa powder** unsweetened, divided	125 mL
3/4 cup	**milk**	175 mL
1/2 cup	**butter or margarine** melted	125 mL
1 tsp	**vanilla extract**	5 mL
1/2 cup	**brown sugar** packed	125 mL
1 cup	**water**	250 mL
4 cups	**strawberries** fresh	1000 mL
	vanilla or coffee ice cream	

Chocolate-Strawberry Port Cake

Chocolate-Strawberry Port Cake

Preheat oven to 325°F (160°C). Grease and flour 9-inch (22 cm) round layer cake pan. In double boiler over simmering water, melt butter and 1/2 cup (125 mL) of the chocolate and 1/3 cup (75 mL) of the port; stir and cool.

In mixer bowl beat egg yolks with 6 tablespoons (90 mL) of the sugar until thick and pale. Gradually beat in chocolate mixture; mix in flour and remaining chocolate pieces.

In another bowl, beat egg whites with remaining sugar just until stiff; gradually fold into chocolate batter to blend. Pour into prepared pan. Bake 25 to 30 minutes until pick inserted into center comes out clean. Cool in pan 5 minutes; loosen and invert onto plate.

With back of spoon, press shallow indentation into center of cake. Toss strawberries with remaining port; spoon into center of cake. Pipe or spoon whipped cream around edge.

SERVES 6

1/3 cup	**butter** or **margarine**	75 mL
1 cup	**semisweet chocolate pieces** divided	250 mL
1/2 cup	**port wine** divided	125 mL
2	**eggs** separated	2
1/2 cup	**sugar** divided	125 mL
1/2 cup	**flour**	125 mL
4 cups	**strawberries** sliced	1000 mL
1 cup	**whipping cream** whipped, sweetened to taste	250 mL

The Ultimate Strawberry Pie

Combine piecrust mix and almonds; mix in water as directed on package. Roll out pastry, crimp edges of shell. Prick bottom.

Bake in 400°F (200°C) oven 8–10 minutes, until golden brown. Set aside to cool. In mixing bowl beat cream cheese and sugar until smooth. Gradually beat in 1/2 of the berries and crush. Spread cream cheese mixture in pastry shell. Dip tips of remaining berries in melted chocolate; arrange upright over cheese layer. Sprinkle with ground almonds.

SERVES 6

1	**pie crust mix** for single crust 9-inch pie	1
1/2 cup	**almonds** finely ground	125 mL
	water	
1 cup	**cream cheese** softened	250 mL
1-1/2 tbsp	**sugar**	20 mL
4 cups	**strawberries** fresh, divided	1000 mL
2 oz	**chocolate** semi-sweet, melted	57 g
	ground almonds for garnish	

Carol Ann Shipman **Berries Cookbook**

Strawberry Shortcake

Strawberry Shortcake

In a large bowl, mix strawberries with Kirsch, Triple Sec, honey and orange peel. Cover and chill for 4 hours, stirring occasionally. In individual serving dishes, arrange 6–8 cubes of cake; top with strawberries. Garnish with whipped cream and a dusting of confectioners' sugar.

SERVES 8

1	**prepared pound cake** or angel-food cake cut into 1-inch cubes	1
6 cups	**strawberries** sliced	1500 mL
3 tbsp	**Kirsch**	45 mL
3 tbsp	**Triple Sec or orange liqueur**	45 mL
2 tbsp	**honey**	30 mL
2 tsp	**grated orange peel**	10 mL
	whipping cream	
	confectioners' sugar	

Strawberry Sauté

In medium skillet melt butter until bubbly. Add walnuts; cook and stir briefly, just until nuts begin to color. Add brown sugar and stir for less than 30 seconds.

Remove pan from heat. Add strawberries and chocolate. Quickly stir to coat strawberries and melt chocolate. Spoon over ice cream, custard pie or angel food cake. Sprinkle with crumbled cookies.

YIELD 2 3/4 cups

3 tbsp	**butter**	45 mL
1/4 cup	**walnuts** coarsely chopped	60 mL
3 tbsp	**dark brown sugar**	45 mL
2 cups	**strawberries** fresh, halved or quartered	500 mL
3 tbsp	**chocolate** semi-sweet, grated	45 mL
1/4 cup	**crumbled cookies**	60 mL

Strawberries in Balsamic Red Wine Sauce

In a small saucepan combine jam, vinegar and wine. Cook and stir over medium heat until smooth and blended, then simmer until syrupy and reduced by half.

Place berries in serving bowl and pour wine mixture over. Gently toss to coat completely.

Serve with whipped cream, if desired.

SERVES 4

2 tbsp	**raspberry jam** seedless	30 mL
1 tbsp	**balsamic vinegar**	15 mL
1 tbsp	**dry red wine**	15 mL
2 cups	**strawberries** fresh, quartered	500 mL
1/4 cup	**whipping cream** whipped, optional	60 mL

STRAWBERRIES

Little Chocolate Spice Cakes with Strawberries & Warm Chocolate Sauce

SERVES 8 – 12

CAKE

1/3 cup	**chocolate chips** semi-sweet	75 mL
3 tbsp	**butter** unsalted	45 mL
1/4 cup	**sifted flour**	60 mL
1/4 tsp	**cinnamon**	1 mL
	pinch of ginger	
	pinch of nutmeg	
	pinch of cloves	
1/4 tsp	**salt**	1 mL
2	**egg whites** room temperature	2
6 tbsp	**sugar**	90 mL

CHOCOLATE SAUCE

6 tbsp	**milk**	90 mL
1/4 cup	**sugar**	60 mL
1 tsp	**vanilla extract**	5 mL
1/2 tsp	**instant espresso powder**	2 mL
2/3 cup	**chocolate chips** semi-sweet	150 mL

FOR GARNISH

1/4 cup	**whipping cream**	60 mL
2-1/2 cups	**strawberries** sliced fresh	625 mL
8	**strawberries** whole, small	8

Cake

Preheat oven to 375°F (190°C). You need a muffin pan with cups 2-3/4 inches in diameter. Butter and flour tins. Melt chocolate and butter together over hot but not boiling water. Turn off heat and stir until smooth. Allow to cool while proceeding with recipe.

Mix the sifted flour thoroughly with the spices and salt. Set aside.

Whip the egg whites until soft peaks form. Continuing to whip, gradually add the sugar. Continue whipping on high speed another minute. The whites will be stiff and glossy.

Sprinkle the flour-spice mixture over the stiff egg whites a little at a time and slowly mix in to just combine. Stir in the chocolate butter mixture to combine.

Fill each muffin cup half-way and bake in the preheated oven 14 to 15 minutes. The tops will be flat and shiny and the cakes just set. Allow to cool 5 minutes in the pan before removing.

Cool thoroughly and wrap to keep moist until ready to use later that day or the next day. (The cakes also freeze well.)

Chocolate Sauce.

Heat the milk with the sugar, vanilla and espresso powder in a small saucepan until just hot. Remove the pan from the heat and stir in the chocolate to melt. Stir until very smooth. Set aside or refrigerate up to one week and reheat when desired.

STRAWBERRIES

Little Chocolate Spice Cakes with Strawberries & Warm Chocolate Sauce

To Serve

Whip cream till soft but distinct peaks form. Refrigerate up to one hour. If chocolate sauce is cold, gently heat until warm.

Split each cake horizontally and put the bottom of each on individual plates. Spoon a small dollop of whipped cream on each bottom and cover that with a small spoonful of sliced strawberries. Place the cake top on the strawberries.

Spoon warm chocolate sauce over top of the cakes and around the plate. Scatter the remaining sliced strawberries around the cake. Place a small dollop of whipped cream on top of each cake and finish with a whole small strawberry. Serve immediately.

Strawberry Tart

Line a 9 x 2 inch (22 x 5 cm) tart pan, with removable bottom, with pastry. Pierce bottom of crust with a fork. Bake at 400°F (200°C) for 10 minutes or until golden brown. Cool. In a medium bowl, whip cream cheese until light and fluffy. Add liqueur and confectioners' sugar. Mix well. In small saucepan, combine cranberry-strawberry drink, sugar and cornstarch; mix well. Cook and stir until mixture thickens and boils. Spread cream cheese mixture over pastry, arrange strawberries on top, stemmed side down, and drizzle with cranberry-strawberry glaze. Refrigerate at least 1 hour.

SERVES 8

pastry for 9-inch single pie crust		
8 oz	**cream cheese** softened	250 g
2 tbsp	**almond-flavored liqueur** or 1/2 tsp (2 mL) almond extract	30 mL
2 tbsp	**confectioners' sugar**	30 mL
1/2 cup	**cranberry-strawberry drink**	125 mL
1 tbsp	**sugar**	15 mL
2 tsp	**cornstarch**	10 mL
4 cups	**strawberries** fresh, hulled	1000 mL

Carol Ann Shipman **Berries Cookbook**

Strawberry Pops

Strawberry Pops

In blender container, blend all ingredients about 1 minute until smooth. Pour into eight 3-ounce, wax-coated paper cups. Place in shallow pan and insert a wooden craft stick or plastic spoon into the center of each. Freeze until firm, about 4 hours.

After pops are frozen, they can be transferred to a resealable plastic bag for freezer storage. To release pops from cups dip briefly into hot water up to rim of cup.

SERVES 8

2 cups	**strawberries**	500 mL
5 oz can	**evaporated milk**	150 mL
3 tbsp	**frozen orange, cranberry or pineapple juice concentrate**	45 mL

Chocolate Strawberry Ice Cream Cake

Fill 8-inch (20 cm) springform pan with ice creams, alternating flavors to create a marbled effect. Swirl in Strawberry Sauce. Smooth surface; freeze until firm. Frost with Chocolate Fudge Sauce. Return to freezer until sauce is firm. To serve, remove from pan about 15 minutes before serving.

Strawberry Sauce
Slightly thaw strawberries; pour into saucepan. Mix sugar and cornstarch. Stir into strawberries. Bring to boil. Cook and stir 2 minutes, or until sauce is slightly thick. Remove from heat. Add lemon juice. Chill.

Chocolate Fudge Sauce
In heavy saucepan, combine whipping cream, sugar and unsweetened chocolate. Melt over low heat, stirring constantly. Stir in rum, if desired. Cool.

SERVES 6 – 8

ICE CREAM CAKE

4 cups	**vanilla ice cream** softened	1000 mL
4 cups	**pistachio ice cream** softened	1000 mL
	strawberry sauce (recipe follows)	
	chocolate fudge sauce (recipe follows)	

STRAWBERRY SAUCE

10 oz	**strawberries** frozen, sliced, in syrup	300 mL
2 tbsp	**sugar**	30 mL
1/2 tbsp	**cornstarch**	7.5 mL
2 tsp	**lemon juice**	10 mL

CHOCOLATE FUDGE SAUCE

1/4 cup	**whipping cream**	60 mL
1/2 cup	**sugar**	125 mL
2 oz	**chocolate** unsweetened	57 g
1 tbsp	**rum**	15 mL

blackberries

To spot the finest blackberries use your instincts. Look for berries that look ALIVE, moist and fresh. They shouldn't be squeezed, flattened or dripping juice. If we say blackberries, we bet you'll answer, "Pie!" or other family favorites. Take home hand-picked blackberries and taste the finest berries in the world on sundaes, cereal, fruit, salads and more.

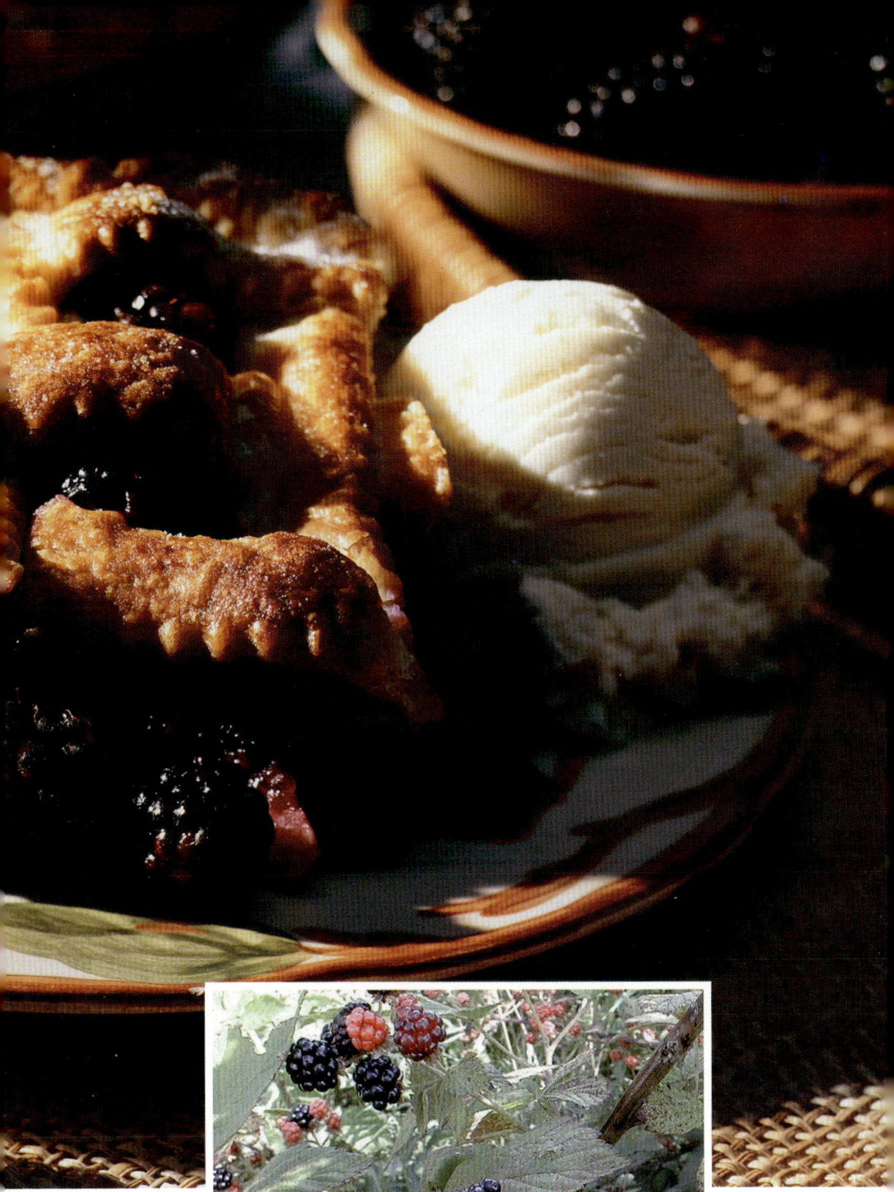

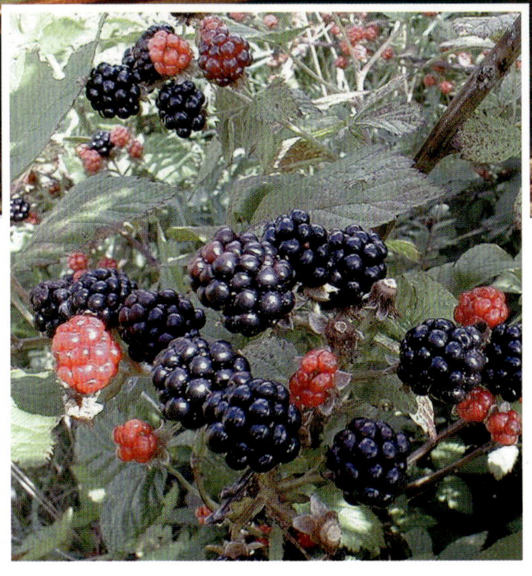

Wild Berry Shake

SERVES 4

1 cup	**blackberries or blueberries**	250 mL
1 cup	**raspberries or strawberries** (with all the juices)	250 mL
1 cup	**black cherry yogurt**	250 mL
	Sprite to taste	
1 scoop	**vanilla ice cream**	1 scoop
	dash of raspberry liqueur and Triple Sec	

Frost standard milkshake glasses or other large glasses in freezer until ready to serve.

Combine all ingredients together in a blender until smooth and creamy.

Garnish with whole berries.

Berry Cool Margaritas

SERVES 2

1 cup	**blackberries** including juice	250 mL
1/4 cup	**tequila**	60 mL
2 tbsp	**orange liqueur**	30 mL
1/4 cup	**lime juice**	60 mL
2 cups	**crushed ice**	500 mL

Blend blackberries and juice in blender for 10–12 seconds. Add remaining ingredients and blend until ice is almost smooth. Moisten rim of tall glasses with lime or orange and dip in sugar. Pour blended mixture into glasses.

Pour puréed blackberries through sieve before adding remaining ingredients if you prefer a seedless drink.

For a non-alcoholic drink omit tequila and orange liqueur and add 1/2 cup (125 mL) of orange juice.

Blackberry Black

SERVES 2

2 oz	**light rum**	60 mL
2 oz	**apple juice**	60 mL
1/2 oz	**lime juice**	15 mL
1/2 cup	**blackberries**	125 mL
1/2 tsp	**superfine sugar**	2 mL
1	**banana**	1

Combine all ingredients in a blender and blend well at high speed. Pour into a 10-ounce (300 mL) Collins glass.

Wild Berry Shake

BLACKBERRIES

The Inn on the Common Wild Blackberry Cordial

SERVES 4 –6

2 cups	blackberries	500 mL
1 cup	sugar	250 mL
2 cups	vodka	500 mL

Mash the blackberries and place them in a large, clean jar. Add the sugar and vodka and stir until the sugar is dissolved. Cover and store in a cool, dark place for 8 weeks, stirring and/or shaking once a week. At the end of the 8 weeks, strain through cheesecloth or a coffee filter into a clean bottle.

Anthony Pace's Butterscotch Blackberries

SERVES 4

3/4 cup	blackberries	175 mL
2 tbsp	butter	30 mL
2 tbsp	scotch whiskey	30 mL
2 tbsp	brown sugar	30 mL
	ice cream, french toast or angel food cake	

In skillet over medium-high heat, melt butter and cook just until beginning to brown slightly.

Add blackberries, cook and stir just until they begin to give off juice. Add the scotch and then ignite with long-handled match. Stir until flames die. Stir in brown sugar and cook until dissolved and sauce is thick. Serve at once over ice cream, french toast or angel food cake.

Blackberry Eggnog

SERVES 10

1 cup	milk	250 mL
4 cups	eggnog	1000 mL
1/2 cup	blackberries	125 mL
1 cup	whipping cream whipped	250 mL

Combine eggnog, milk and blackberries. Beat until frothy. Top with real whipped cream.

Carol Ann Shipman **Berries Cookbook**

BLACKBERRIES

Chef Martha Crawford's Double Berry Super Sodas

Strawberry Purée

Place strawberries in blender or food processor and purée until smooth. Strain and refrigerate until ready to make sodas.

Frost 4 to 6 very tall glasses as follows: spread 1/4 (60 mL) cup sugar evenly on small plate, measure 1/4 cup (60mL) water into small bowl; dip rims of glasses first in water, then in sugar to coat rims; keep glasses in freezer until ready to prepare sodas.

For Each Soda

Layer berries and ice cream in frosted glasses; with 1/4 of the berries, small scoop ice cream, another 1/4 of the berries, second scoop of ice cream, another 1/4 of the berries. Sodas can be held in freezer for several hours at this point.

In large pitcher, gently blend purée, sparking water, lemon juice and brandy. Pour into prepared frosted glasses.

Garnish with mint leaves and remaining berries.

SERVES 4 – 6

1 lb	**strawberries** for purée	454 g
1 lb	**strawberries blackberries or blueberries** (for sodas)	454 g
2 cups	**vanilla ice cream** or frozen yogurt	500 mL
3	**sparkling water or seltzer** 12-ounce cans, chilled	3
1 tbsp	**lemon juice**	15 mL
1 tbsp	**cherry brandy** if desired	15 mL

Blackberry Summer Tea

SERVES 2 – 4

1 cup	boiling water	250 mL
7	blackberry-flavor tea bags	7
1 cup	pineapple juice	250 mL
10 oz	crushed pineapple undrained	300 mL
1/3 cup	sugar	75 mL
	few whole blackberries for garnish	
	ice cubes	

Pour boiling water over tea bags; steep for 10 minutes.

Discard tea bags and chill. Combine in blender all ingredients.

Blend on high speed until smooth. Crush ice in a clean tea towel and add to blender. Serve in tall chilled frosted glasses with a straw and float a couple whole blackberries on top for decoration.

Blackberry Smoothie

SERVES 4 – 6

1 cup	orange juice	250 mL
1-1/2 cups	peach nectar	375 mL
1-1/2 cups	plain yogurt	375 mL
1	banana	1
1 cup	blackberries frozen	250 mL

Combine orange juice, peach nectar, and yogurt together. Slice banana in large chunks. Keep blackberries frozen until ten minutes before adding to the blender. On high speed, blend banana and frozen blackberries; add remaining ingredients, blend until creamy. Serve in tall and frosted glasses.

Blackberry Summer Salad

SERVES 6

2-1/2 cups	blackberries canned or frozen in syrup	625 mL
2	nectarines or peaches medium size, pared and sliced	2
2	kiwi fruit medium size, peeled and sliced or	2
6	pineapple spears	6
1/2	cantaloupe or honeydew melon medium size, cut into spears	1/2
2	bananas medium size, peeled and sliced	2
6	lettuce leaves	6
	blackberry yogurt dressing	

Drain blackberries, reserving syrup for dressing. Arrange fruit on lettuce-lined salad plates. Serve blackberry yogurt dressing with salad.

Blackberry Yogurt Dressing

Stir 1/4 cup (60 mL) reserved blackberry syrup, 1 tablespoon (15 mL) honey and 2 teaspoons (10 mL) chopped mint into 1 (8 ounces) carton low-fat plain yogurt.

Fruit may be cut into bite-size pieces and mixed with Blackberry Yogurt Dressing and served in glass bowl.

Oregon Blackberry Soup

Sprinkle blackberries with lemon juice and sugar, toss and set aside for 45 minutes. In small bowl combine preserves and cream, whisk until well blended. Stir in wine or grape juice and berries, including any juice. Cover and refrigerate for 3 hours. Using a ice cream scoop, make balls with lemon ice cream or sherbet for topping of soup. Set aside.

Ladle cold soup into individual bowls or stemmed glasses. Top each with 4 ice cream balls.

SERVES 6

2 cups	**blackberries** fresh or frozen unsweetened	500 mL
1 tbsp	**lemon juice**	15 mL
1/4 cup	**sugar**	60 mL
1/2 cup	**blackberry preserves**	125 mL
1 cup	**half and half cream**	250 mL
1/2 cup	**dry white wine or white grape juice**	125 mL
2 cups	**lemon ice cream or sherbet**	500 mL

Windsong Mountain Inn Blackberry Soup

In a medium, non-aluminum saucepan combine 2-1/2 cups (625 mL) of blackberries and water. Bring to a simmer, cover and cook 15 minutes or until blackberries are very soft. Press blackberry mixture through a sieve over a small bowl, reserving the liquid. Discard the seeds left in the sieve. Combine blackberry liquid, 3 tablespoons (45 mL) of sugar and Kirsch. Cover and chill mixture for at least 2 hours.

Combine 3 tablespoons (45 mL) of sugar and peaches. Toss gently to coat. Spoon 1/4 cup blackberry mixture into each of 6 shallow fruit or soup bowls. Arrange 1/3 cup (75 mL) peach slices and remaining blackberries over each serving.

SERVES 6

3-1/2 cups	**blackberries** divided	875 mL
1/2 cup	**water**	125 mL
6 tbsp	**sugar** divided	90 mL
1 tsp	**Kirsch or cherry brandy**	5 mL
2 cups	**peaches** sliced, peeled	500 mL

Carol Ann Shipman **Berries Cookbook**

Salmon with Blackberry Hollandaise

SERVES 4

HOLLANDAISE

4	**egg yolks**	4
2 tbsp	**lemon juice** fresh	30 mL
1/4 cup	**butter** melted	60 mL
	dash salt, white pepper	

BLACKBERRY SAUCE

2 cups	**blackberries**	500 mL
1 tsp	**dijon mustard**	5 mL
1/2 tsp	**garlic** minced	2 mL
1/2 tsp	**shallot** minced	2 mL
1	**lemon** juiced	1
3 tbsp	**Madeira wine**	45 mL

SALMON

1	**lemon** quartered	1
	fresh mint	
2 cups	**white wine**	500 mL
3 cups	**fish stock**	750 mL
4	**salmon fillets** 6 oz (170 g) each	4

Hollandaise

Whisk egg yolks and lemon juice in top pan of double boiler. (Water in bottom of double boiler should be simmering, and not touching top pan). Place the pan onto double boiler. Whisk continuously until sauce just begins to thicken.

Remove from heat and slowly pour in melted butter in steady stream, continuing to whisk until all butter is incorporated. Add salt and white pepper. Set aside.

Blackberry Sauce

In small saucepan, combine blackberries (reserving 12–20 whole blackberries for garnish), mustard, garlic, shallot, lemon juice and Madeira. Reduce by one-fourth, stirring often to keep from scorching. Purée in a food processor; pass through a fine sieve. Set aside one quarter of the reduction for garnish. Fold the remaining portion into the hollandaise.

In a fresh saucepan, combine quartered lemon and one sprig of mint with white wine and fish stock. Bring to a boil, then reduce to simmer. Using a spatula, carefully place the salmon fillets in saucepan, and poach for 8–10 minutes.

Remove fillets with spatula so that they retain their shape. Spoon some Blackberry Hollandaise into the center of each serving plate, and then drizzle a thin circular line of reserved purée. Lightly draw the back of a spoon across it, making a scallop effect toward the outside of the plate. Place a salmon fillet in the circle and garnish with mint leaves and 3–5 whole blackberries.

Beef Tenderloin with Blackberry Port Wine

In a saucepan bring diced shallot, 3/4 cup (175 mL) blackberries, wine and sugar to boil. Boil gently to reduce wine to 1/2 cup (125 mL). Strain and set liquid aside. Boil beef stock in separate pan to reduce by half. This will take approximately 15 minutes. Grill steaks or pan broil in a skillet 3 to 4 minutes per side. Whisk blackberry and port wine reduction into reduced beef stock.

If sauce is too thin, dissolve 1 teaspoon (5 mL) of cornstarch in water, then stir into sauce and bring to boil. Whisk in 1 tablespoon (15 mL) softened butter. Serve steaks with sauce and garnish with remaining blackberries.

SERVES 4

1	**large shallot** or small onion, finely diced	1
1 cup	**blackberries** fresh or frozen, divided	250 mL
2 cups	**port wine**	500 mL
1 tsp	**sugar**	5 mL
2 cups	**beef stock**	500 mL
1 tbsp	**butter** softened	15 mL

Blackberry Sauce

Combine blackberries, sugar and lemon juice in a small saucepan. Cover and cook until bubbling. Remove from the heat, place in a food processor and blend. Pass through a strainer to remove the seeds.

Chill before serving. Serve with duck.

YIELD 2 cups

2 cups	**blackberries**	500 mL
3 tbsp	**sugar**	45 mL
1 tbsp	**lemon juice**	15 mL

Drying Firm Blackberries

Crack the skins by dipping the blackberries in rapidly boiling water for 15 to 30 seconds, then plunge them into cold water. Remove the excess moisture and dry in a dehydrator for 12 to 24 hours. They have dried properly if they are leathery, pliable and contain no moisture when crushed.

Pork Chops and Blackberry Sauce

SERVES 2 – 4

4	**pork chops** center cut	4
	oil small amount for pan-frying	
2 tbsp	**butter**	30 mL
1/4 cup	**yellow onion** chopped	60 mL
1/2 cup	**sherry**	125 mL
1/3 cup	**blackberry purée**	75 mL
1/4 cup	**currant jelly**	60 mL
1/4 cup	**chicken stock**	60 mL
1 tbsp	**cornstarch**	15 mL
1/2 cup	**blackberries**	125 mL

Add small amount of oil to pan. Brown pork chops on both sides; reduce heat and continue cooking until no pink remains. Remove to platter and keep warm.

Melt butter in saucepan. Add onion and sauté until transparent. Add sherry and simmer until reduced by 1/3. Mix unsweetened blackberry purée, currant jelly, chicken stock and cornstarch together. Add to hot sherry in small amounts, mixing and stirring until thickened. Remove from heat and gently fold in blackberries. Pour over pork chops to serve.

This sauce is equally good over braised chicken breasts or curried chicken.

Blackberry Purée
Place fresh or thawed whole frozen blackberries in food processor and process until smooth. If desired, seeds may be removed by straining through a medium sieve and using a rubber spatula to press pulp through while scraping underside of the sieve. Add sugar to taste.

A good rule of thumb for sweetening is about 2 tablespoons (30 mL) sugar per cup of whole berries.

Crab Wontons with Blackberry Szechuan Sauce

Blackberry Szechuan Sauce
Mix all ingredients in saucepan. Bring to a boil over medium-high heat and cook until clear and thickened. The flavor of this sauce improves after standing overnight.

Filling
Wash spinach. With water still clinging to leaves, place in large pan over medium-high heat. Cook until spinach just begins to wilt and most of water has evaporated. Empty onto cutting board and chop finely. Set aside.

Melt butter in sauté pan. Add onion and sauté until transparent. Reduce heat to low; add cream cheese. When the cheese begins to soften, add lemon juice to blend. Remove from heat and stir in crab, breadcrumbs and spinach.

Wontons
Place 1–2 teaspoons (5–10 mL) of filling on each wrapper and seal. Place single layer of wontons in hot oil and fry 2–3 minutes until golden brown. Drain on paper towels and serve immediately with Blackberry Szechuan Sauce.

YIELD about 3 dozen

SAUCE

1/2 cup	blackberry purée	125 mL
1/2 cup	sake or dry sherry	125 mL
1 tbsp	cornstarch	15 mL
1/2 tsp	salt	2 mL
1/2 tsp	red pepper flakes	2 mL
1/2 tsp	ginger grated	2 mL
1 tsp	lime juice	5 mL
2	cloves garlic minced	2
1-1/2 tbsp	honey	25 mL

FILLING

2–3 oz	fresh spinach trimmed	55–85 g
1 tbsp	butter	15 mL
4 tbsp	onion chopped finely	60 mL
3 oz	cream cheese cut into small chunks.	85 g
2 tbsp	dry breadcrumbs	30 mL
1/2 lb	cooked crabmeat flaked	227 g
	dash salt, pepper and **tabasco** (optional)	

WONTONS

approximately 3 dozen wonton wrappers

vegetable oil to cover bottom of wok to 1/4 inch.

Raspberries & Blackberries with Champagne Cheeses

SERVES 4 – 6

4 oz	**goat cheese**	125 g
4 oz	**cream cheese**	125 g
2 tbsp	**pine nuts** coarsely chopped	30 mL
1 tbsp	**fine dry bread crumbs**	15 mL
2 tbsp	**sugar**	30 mL
2 tbsp	**champagne vinegar**	30 mL
1 tbsp	**olive oil**	15 mL
1 tbsp	**honey**	15 mL
	pinch each black pepper, garlic powder, salt	
6 oz	**raspberries**	170 g
6 oz	**blackberries**	170 g

Preheat oven to 400°F (205°C). Mix goat cheese and cream cheese in food processor or by hand until blended. Form mixture into a round ball about the size of a hockey puck. Place on oven-proof plate or an aluminum foil coated cookie sheet. Combine chopped pine nuts and bread crumbs in a small bowl. Press onto top and sides of cheese ball. Place cheese ball in center of oven for 10 minutes or until lightly browned. In a medium bowl, combine sugar, vinegar, olive oil, honey, pepper, garlic and salt. Gently mix in berries. Flatten top of cheese ball, spoon berries over top. Serve warm with flat bread, crackers or French bread.

Raspberries & Blackberries with Champagne Cheeses

Berry Empanadas

Preheat oven to 400°F (200°C). Combine apple, walnuts, sugar, flour, cinnamon, vanilla and salt in medium bowl and mix well. Gently fold in berries. Set aside.

Roll out pastry and cut into twelve 4-1/2 inch rounds. Place 2 tablespoons (30 mL) berry filling on half of each round, leaving about 1/2 inch along edges. Fold the other half of pastry over filled half; moisten edges and seal by pressing with fork.

Repeat this procedure until all pastry rounds are folded and sealed.

Combine 2 tablespoons (30 mL) sugar with 1/2 teaspoon (2 mL) ground cinnamon. Brush each empanada with melted butter and sprinkle with cinnamon-sugar. Bake on greased cookie sheet 18–20 minutes or until golden brown.

YIELD 16 – 24

2 cups	**blackberries** raspberries or boysenberries	500 mL
1 cup	**peeled apple** finely chopped about 1 medium size, tart variety such as Granny Smith	250 mL
1/4 cup	**chopped walnuts**	60 mL
1/4 cup	**sugar**	60 mL
2 tbsp	**flour**	30 mL
1 tsp	**ground cinnamon**	5 mL
1 tsp	**vanilla**	5 mL
	dash of salt	
3	**single pastry crusts** for 9 inch pie	3
1 tbsp	**butter**	15 mL
	cinnamon-sugar for dusting	

Grandma's Blackberry Cake

Toss blackberries with 3 tablespoons (45 mL) of flour; set aside. In a mixing bowl cream butter and sugar. Add eggs; beat well.

Combine baking soda, cinnamon, nutmeg, salt, cloves, allspice and remaining flour; add to cream mixture alternately with buttermilk.

Fold in blackberries, pour into a greased and floured 9 x 9 inch (22 x 22 cm) baking pan. Bake at 350°F (175°C) for 45 to 50 minutes or until cake tests done. Cool on wire rack.

Serve with whipped cream.

SERVES 6 – 8

1 cup	**fresh blackberries**	250 mL
1 tsp	**ground cinnamon**	5 mL
2 cups	**all-purpose flour** divided	500 mL
1 tsp	**ground nutmeg**	5 mL
1/2 cup	**butter or margarine** softened	125 mL
1/2 tsp	**salt**	2 mL
1 cup	**sugar**	250 mL
1/4 tsp	**ground cloves**	1 mL
2	**eggs**	2
1/4 tsp	**ground allspice**	1 mL
1 tsp	**baking soda**	5 mL
3/4 cup	**buttermilk**	175 mL
	whipped cream optional	

Scrumptious Berry Torte

SERVES 10

2	**puff pastry sheets**	2
4 cups	**blackberries** or **fresh blueberries**	1000 mL
4 cups	**strawberries** fresh	1000 mL
4 cups	**whipping cream** whipped	1000 mL

Bake puff pastry sheets according to package directions. Whip cream and set aside for assembly of torte. Cut the puff pastry sheets lengthwise in half.

Mash some of the strawberries and place on bottom layer. Next put a layer of whipping cream and then blackberries or blueberries.

Continue with the next layer of puff pastry whipping cream and berries. Arrange a layer of both berries on the top in your own design.

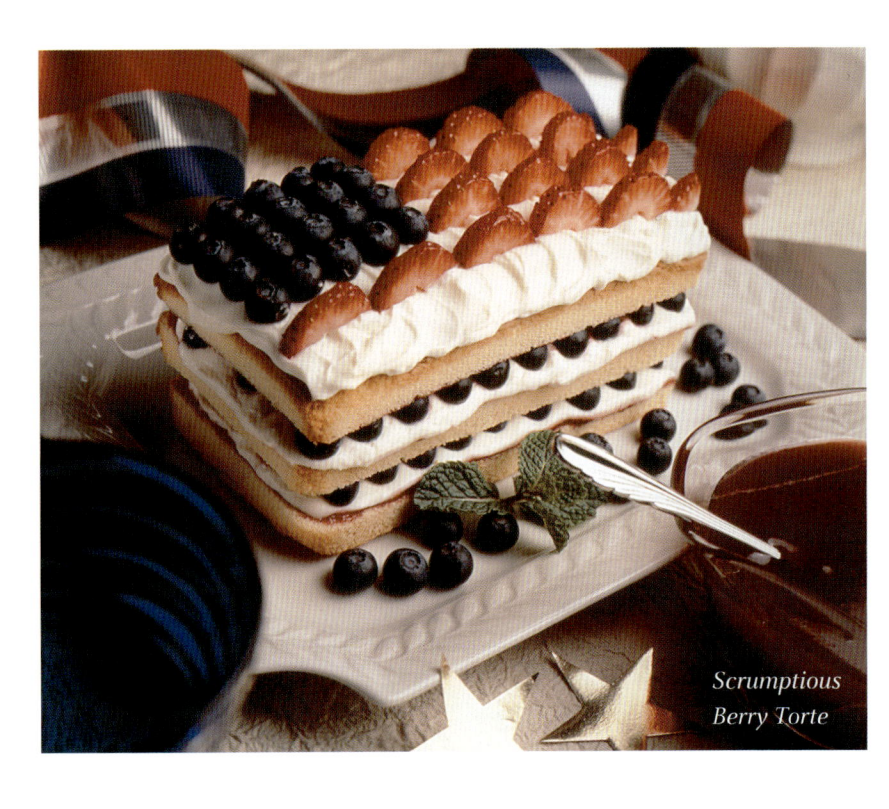

Scrumptious Berry Torte

Blue Spruce Inn Fruit Pizza Lorenzi

Preheat oven to 400°F (200°C). Crust is best made with a food processor.

Crust
Place egg, margarine, shortening, salt, cream of tartar, flour, baking soda and sugar into food processor and mix just till blended. Press crust into 13-inch pizza pan. Bake for 8–10 minutes.

Cheese Filling
Use the food processor to combine cream cheese, sugar and juice. When the crust is cool, spread with cheese filling.

Topping
Select a variety of fresh fruit to arrange on the top of the pizza. To make a glaze for the fruit, either melt 1/2 cup (125 mL) jelly and brush over fruit or combine leftover fruit juice with 1 teaspoon (5 mL) cornstarch and 1/4 cup (60 mL) sugar. Bring to boil and stir until thick and smooth. Brush over the fruit.

SERVES 10 – 12

1	egg	1
1/4 cup	margarine	60 mL
1/4 cup	shortening	60 mL
3/4 cup	sugar	175 mL
1-1/3 cup	flour	325 mL
1/2 tsp	baking soda	2 mL
	a variety of seasonal fruits blackberries strawberries blueberries	
1/8 tsp	salt	0.5 mL
1 tsp	cream of tartar	5 mL
8 oz	cream cheese	250 g
1/3 cup	sugar	75 mL
2 tsp	fruit juice	10 mL

West Coast Blackberry Pie

Combine berries, sugar, flour, lemon juice and salt. Line large pie plate with pastry, fill with berry mixture; dot with butter and cover with top crust. Bake in very hot oven 450°F (230°C) 10 minutes; reduce temperature to moderate 350°F (175°C) oven and bake 25 to 30 minutes longer or until berries are tender.

SERVES 6

3 cups	blackberries fresh	750 mL
1/8 tsp	salt	0.5 mL
1 cup	sugar	250 mL
2 tbsp	flour	30 mL
1 tbsp	butter	15 mL
2 tbsp	lemon juice	30 mL
1	recipe plain pastry for double-crust pie	1

saskatoon
berries

Luscious sweet Saskatoons invite us to "sit for a minute." In muffins, pies, floating in milk or rolling down a mountain of ice cream, the mouth-watering flavor of Saskatoon berries turns a taste into a summer picnic.

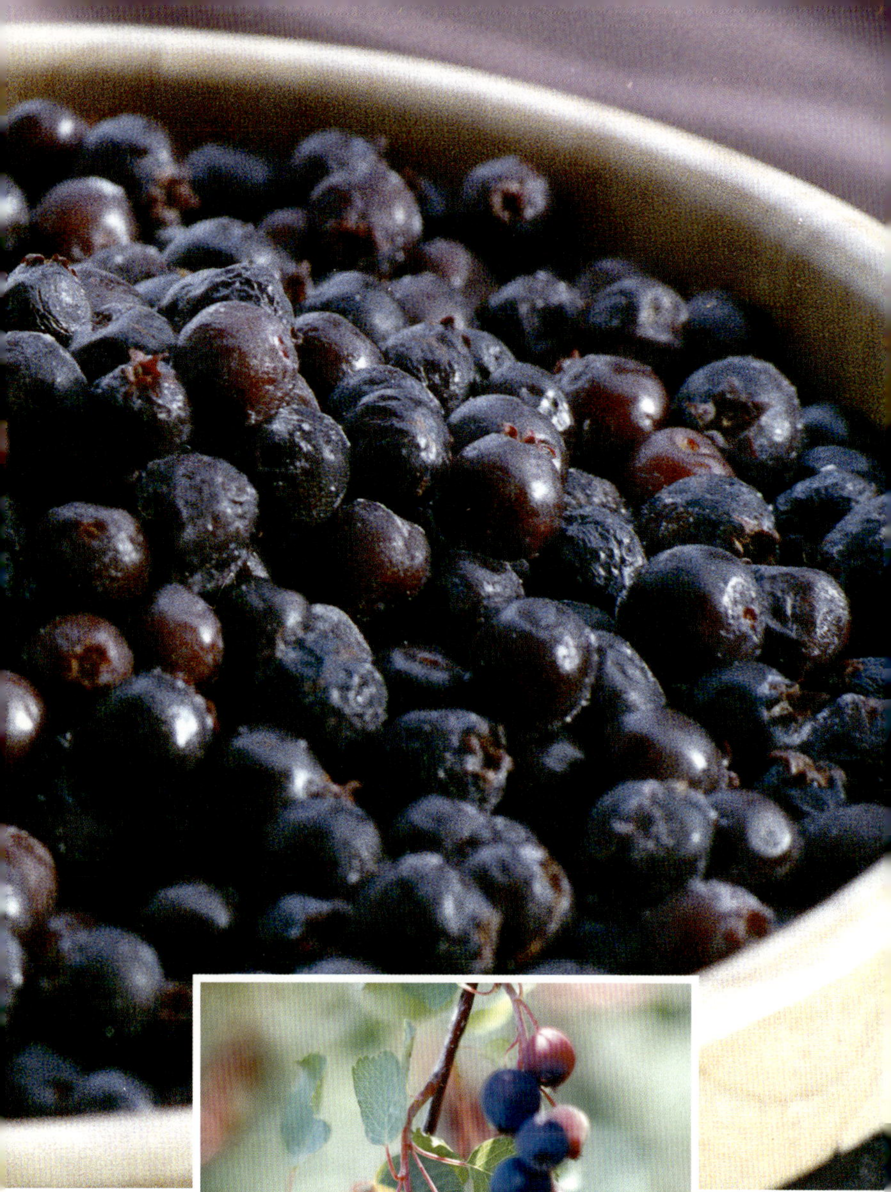
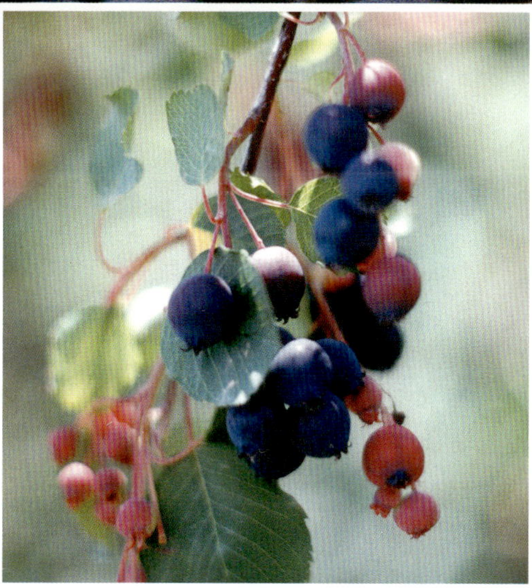

SASKATOON BERRIES

Saskatoon & Tortellini Fruit Salad

SERVES 2

| 1/2 cup | **poppy seed dressing**
 bottled or homemade | 125 mL |

FRUIT SALAD

1-1/4 cups	**three cheese tortellini pasta**	310 mL
1 cup	**Saskatoon berries** fresh	250 mL
1 cup	**strawberries** sliced fresh	250 mL
11 oz. can	**mandarin orange segments** drained	330 mL
3/4 cup	**green grapes**	175 mL

Cook pasta according to directions on package; drain. In a large bowl, add pasta and fruit salad ingredients. Pour dressing over salad and toss lightly; refrigerate until ready to serve. Three cheese tortellini pasta can be found in the refrigerated section of your grocery store. Other fruits such as bananas, peaches, apples, and oranges may be used.

Saskatoon Berry Salsa

YIELD 2 1/2 cups

2	**oranges**	2
2	**lemons**	2
2 cups	**Saskatoon berry purée**	500 mL
1 cup	**port wine**	250 mL
1 tsp	**mirroir topping** dry mix	5 mL

Grate the rinds of the 2 oranges and the 2 lemons. Mix port wine with the grated orange and lemon rind. Reduce for 10 minutes. Squeeze the oranges and lemons and set the juice aside. Strain the 2 cups (500 mL) Saskatoon berry purée and mix with reduced port wine and fruit rind mix. Reduce the mixture for 10 more minutes. Mix juice from the squeezed oranges and lemons with 1 teaspoon (5 mL) mirroir topping dry mix. Add to the port wine, fruit rind mix. Continue to simmer for 5 minutes. Cool before serving.

Cold Saskatoon Berry Soup

SERVES 2

1/2 cup	**Saskatoon berry purée**	125 mL
1/2 cup	**club soda**	125 mL
1/2 cup	**dry white wine**	125 mL
2	**slices of lime**	2

Mix first three ingredients together, strain through a fine sieve. Serve in two martini glasses, garnished with a slice of fresh lime.

Saskatoon Berry Strudel

Press edges of patty shell together; roll out on floured cloth to 18 x 14 inch rectangle. Place pie filling evenly along longest side of rectangle about 4 inches from edge. Without stretching, carefully fold the 4-inch piece of dough over the filling. Pick up the cloth, making the dough roll forward into a tight roll. Seal ends. Place on a foil-lined baking sheet; curve strudel, forming a crescent or horseshoe shape. In a small cup, combine melted butter and ground almonds. Brush almond-butter mixture over the top of strudel. Lightly sprinkle the sugar over the strudel. Bake in 350°F (175°C) oven for 45 to 50 minutes. Remove from sheet; cool on rack.

SERVES 6

3	**patty shells** frozen	3
3 cups	**Saskatoon berry pie filling***	750 mL
3 tbsp	**melted butter**	45 mL
1 tbsp	**ground almonds**	15 mL
	coarse sugar granules	

*can be purchased at supermarkets

Chicken With Saskatoon Berry Sauce

Heat a sauté pan, season chicken with salt and pepper; add oil and butter to hot frying pan. Place chicken in frying pan skin side down. Sauté until lightly browned, about 5 minutes.

Turn chicken; bake in 350°F (175°C) oven for 10 minutes. Remove from oven and take chicken from frying pan—keep warm. Add shallots to the pan and sauté for 1 minute.

Deglaze with Grand Marnier; add Saskatoon berry purée, honey and demi-glace or chicken stock to the pan.

Reduce for 2 to 3 minutes; add chicken. Take two dinner plates, pour sauce on each plate, and cut chicken in a fan design on top of the sauce on each plate.

SERVES 2

2	**boneless chicken breasts**	2
2 tbsp	**vegetable oil**	30 mL
1 tbsp	**butter**	15 mL
	salt and pepper for seasoning	
1 tbsp	**shallots** finely chopped	15 mL
1 tbsp	**Grand Marnier**	15 mL
1/3 cup	**Saskatoon berry purée**	75 mL
1 tbsp	**honey**	15 mL
1/3 cup	**demi-glace or chicken stock**	75 mL

Carol Ann Shipman **Berries Cookbook**

SASKATOON BERRIES

Saskatoon Berry Chutney

Soak dried fruit in hot water for half hour, rinse and strain. Chop all ingredients in a food processor – leave fairly chunky. Mix all together until smooth and keep in a jar for 2 to 3 days for flavor to develop.

YIELD 1 3/4 cups

1/2 cup	**Saskatoon berry purée**	125 mL
3 oz	**sundried cherries**	85 g
2 oz	**sundried cranberries**	55 g
1/4 cup	**red onion** diced small	60 mL
1 tsp	**tomato paste**	5 mL
1 tsp	**ancho chili paste**	5 mL
4	**cloves roasted garlic**	4
1/4 cup	**brown sugar**	60 mL
1 tbsp	**balsamic vinegar**	15 mL

Charcoal Grilled Cornish Hen Marinated in Saskatoon Vinegar Marinade

Boil purée and vinegar, strain and cool to room temperature. In a shallow dish, large enough to hold the hens, combine purée mixture, olive oil, garlic and herbs; set aside.

Cut cornish hens in half, place birds in marinade, coat them evenly. Let them stand at room temperature for 2 hours and refrigerate overnight.

Pat birds dry and gently grill over charcoal until juices run clear.

Garnish and Sauce
Melt butter, add Saskatoon berry purée and bring to boil. Add chicken stock and cook sauce until thick. Season and drizzle over birds on serving platter. Garnish with crème fraiche and chives.

SERVES 12 – 14

6	**cornish hens** 1-1/2 pounds each	6

MARINADE

1 cup	**Saskatoon berry purée**	250 mL
5-6 tbsp	**red wine vinegar** of high quality	75 mL
1/4 cup	**olive oil**	60 mL
1	**bay leaf**	1
1	**twig of thyme**	1
	pinch of minced garlic	

GARNISH & SAUCE

2 tsp	**butter**	10 mL
1/4 cup	**Saskatoon berry purée**	60 mL
1 cup	**crème fraiche** or 1/2 cup whipping cream and 1/2 cup sour cream	250 mL
	fresh cut chives	

Medallions of Venison or Buffalo with Saskatoon Berry Sauce

Season medallions with salt and pepper, both sides. Flour lightly on one side. Heat sauté pan over high heat; add vegetable oil and butter. Place meat flour side down in pan. Sauté medallions, each side for 3 minutes. Remove meat from frying pan and keep warm. Add berries to the frying pan, sauté for 1 minute. Deglaze pan with brandy, add Saskatoon berry purée and demi-glace. Reduce slowly for 3 to 4 minutes. Add medallions to the sauce; simmer lightly for 1 minute longer. Serve medallions 3 to a plate. Lightly spoon sauce over the meat.

SERVES 2

6	medallions of venison or buffalo	6
salt and pepper for seasoning		
flour to coat meat		
1 tbsp	vegetable oil	15 mL
1 tbsp	butter	15 mL
1/2 cup	Saskatoon berries	125 mL
1 tbsp	brandy	15 mL
1/3 cup	Saskatoon berry purée	75 mL
2 tbsp	demi-glace*	30 mL

* Dry mix can be purchased at specialty stores.

A "Canadian" Sauce Cumberland

Blend all ingredients well and refrigerate overnight to allow the flavors to blend. Use sauce for game or poultry.

YIELD 2 1/2 cups

2-3/4 cups	Saskatoon berry purée	675 mL
1-1/4 cups	royal red currant purée*	300 mL
1/4 cup	orange zest	60 mL
cayenne pepper to taste		

* Can be purchased at specialty stores. Can be substituted with red currant jam.

SASKATOON BERRIES

Chicken Grill with Tangy Saskatoon Sauce

SERVES 4

3 tbsp	olive oil	45 mL
3 tbsp	**blueberry** or **raspberry vinegar**	45 mL
1-1/2 tsp	**lime juice**	7 mL
2	**garlic cloves** minced	2
4	**chicken breasts** skin removed	4
1 cup	**Saskatoon berries**	250 mL
1 cup	**puréed raspberries**	250 mL

In sturdy plastic zipper-type bag, combine oil, vinegar, lime juice and garlic. Add chicken, turning to coat all sides. Marinate, in refrigerator, 1 to 2 hours. Reserving marinade, remove chicken and pat dry with a paper towel.

Prepare sauce by combining marinade, Saskatoon berries and raspberry purée in a saucepan. Stirring occasionally, place over medium heat and cook 5–7 minutes until slightly thickened. Remove from heat and set aside.

Grill chicken over medium heat, cooking 8–12 minutes per side or until the juices are no longer pink when the chicken is cut. Serve the chicken with the Saskatoon-raspberry sauce. Chicken can also be prepared using the broiler, microwave or pan sauté methods.

Saskatoon Prairie Berry Custard

SERVES 4

3 tbsp	**butter**	45 mL
8	**eggs**	8
1/4 tsp	**salt**	1 mL
2/3 cups	**flour**	150 mL
1 cup	**Saskatoon berries** fresh or frozen	250 mL
1/4 cup	**honey**	60 mL
2-1/2 cups	**milk**	625 mL
1 tsp	**vanilla extract**	5 mL
	dash of nutmeg and confectioners' sugar	

Melt butter in a large 9 x 13-inch (22 x 33 cm) baking dish. Blend all ingredients, except Saskatoon berries, in blender. Pour into preheated baking dish with melted butter and sprinkle with Saskatoon berries. Bake at 425°F (220°C) for 20 to 25 minutes until puffed and golden. Drizzle with hot jam or maple syrup. Sprinkle with nutmeg and confectioners' sugar. In the summer, garnish with fresh raspberries and mint from the garden.

Carol Ann Shipman **Berries Cookbook**

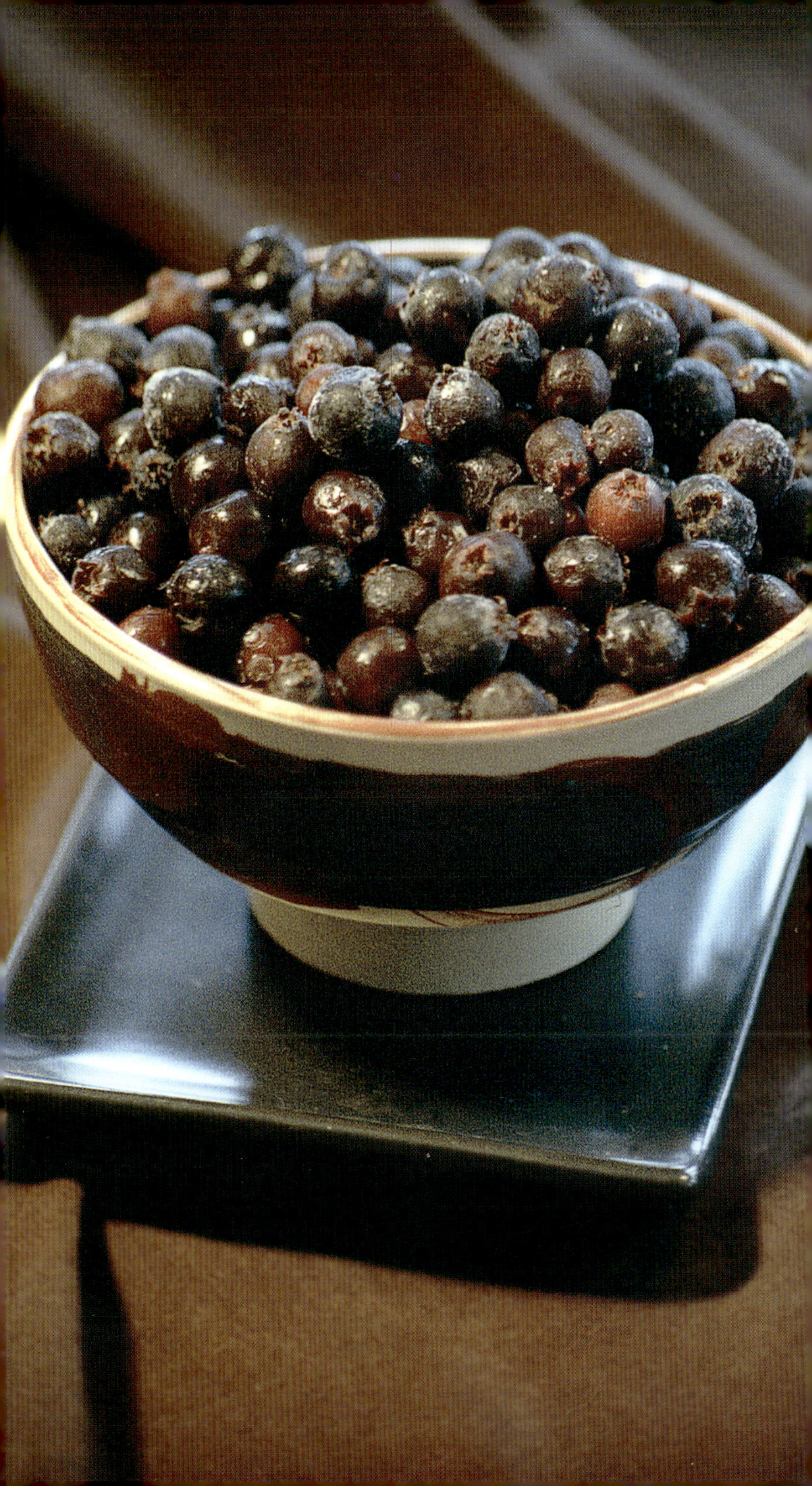

Saskatoon Berry Relish

SERVES 12

1 tbsp	chopped garlic	15 mL
1 tbsp	chopped ginger	15 mL
1/2 cup	butter	125 mL
2 cups	brown sugar	500 mL
2 cups	mango purée	500 mL
2 cups	chicken stock	500 mL
1/2 cup	each diced red and green pepper tomato red onion	125 mL
1 cup	Saskatoon berries	250 mL
	cornstarch to thicken	
	salt and pepper	

Sauté garlic and ginger in butter, then add brown sugar and allow to caramelize slightly. Add mango and chicken stock; reduce until it forms a light syrup.

Sauté all vegetables briefly and add to the sauce along with the berries. Simmer for 10 minutes. Add cornstarch as needed to keep consistency. Salt and pepper to taste.

Saskatoon Crisp

SERVES 6

FILLING

4 cups	Saskatoon berries	1000 mL
1/4 cup	sugar	60 mL
1/2 tsp	lemon rind grated	2 mL
1 cup	apples peeled, diced	250 mL

CRISP

1/2 cup	light brown sugar	125 mL
2 tsp	cinnamon	10 mL
1 tsp	nutmeg	5 mL
1/2 cup	flour	125 mL
1/2 cup	chopped pecans	125 mL
1/2 cup	rolled oats	125 mL
1/8 tsp	salt	0.5 mL
3 tbsp	butter	45 mL

Preheat oven to 325°F (160°C).

Filling
In a small bowl, combine Saskatoon berries, sugar, lemon rind, and apples. Mix well and place in a well-buttered 8 x 8 inch (20 x 20 cm) pan.

Crisp
In a medium bowl, combine brown sugar, cinnamon, nutmeg, flour, pecans, oats and salt; cut in the butter with a pastry cutter until it resembles coarse crumbs. Spread evenly over the Saskatoon berry filling. Bake 45 minutes or until the crust is golden brown.

Saskatoon Berry Yogurt

SERVES 1 – 2

1 cup	plain yogurt	250 mL
3 tbsp	Saskatoon berry syrup	45 mL

Add the syrup to the yogurt and stir well.

Saskatoon-Lemon Cheesecake Bars

Preheat oven to 350°F (175°C). Lightly grease the bottom of a 9 x 13-inch (22 x 33 cm) pan.

In a small saucepan combine Saskatoon filling ingredients; stir until cornstarch is dissolved. Cook over medium heat, stirring constantly until thick and bubbly, about 5 minutes. Set aside to cool slightly.

Meanwhile, in a large bowl, combine all crumb mixture ingredients except butter; mix well. Using a pastry blender or fork, cut in butter until mixture resembles coarse crumbs. Reserve 1 cup (250 mL) crumb mixture for topping. Press remaining crumb mixture firmly in bottom of greased pan. Bake for 10 minutes.

While crust is baking, in a medium bowl mix together cheesecake filling ingredients with electric mixer on medium speed until well blended; pour into baked crust. Spoon Saskatoon filling over cheesecake filling, swirl with knife to blend. Sprinkle reserved crumb mixture over filling. Bake 20 to 25 minutes at 350°F (175°C) until lightly browned. Cool completely. Cut into bars.

SERVES 4

SASKATOON FILLING

1 cup	**Saskatoon berries** chopped	250 mL
1/4 cup	**orange juice**	60 ml
2 tbsp	**sugar**	30 mL
2 tsp	**cornstarch**	10 mL

CRUMB MIXTURE

1-1/4 cups	**flour**	310 mL
3/4 cup	**rolled oats**	175 mL
3/4 cup	**brown sugar** firmly packed	175 mL
1/2 cup	**chopped nuts**	125 mL
1/2 cup	**butter**	125 mL

CHEESECAKE FILLING:

8 oz.	**Neufchatel cheese** softened	250 g
2	**eggs**	2
1/2 cup	**sugar**	125 mL
2 tbsp	**fresh lemon juice**	30 mL
1 tsp	**lemon peel** grated	5 mL

Saskatoon Pie

In a saucepan, simmer Saskatoon berries in water for 10 minutes. Add lemon juice. Stir in sugar mixed with flour. Pour into pastry-lined pie plate. Cover with top crust; seal edges. Bake in 425°F (220°C) oven for 15 minutes; reduce heat to 350°F (175°C) oven and bake 35 to 45 minutes longer or until golden brown.

SERVES 6

4 cups	**Saskatoon berries**	1000 mL
1/4 cup	**water**	60 mL
3/4 cup	**sugar**	175 mL
3 tbsp	**flour**	45 mL
2 tbsp	**lemon juice**	30 mL
	pastry for 1 double-crust pie	

SASKATOON BERRIES

Saskatoon Granola Bars

YIELD 12 – 16 bars

1/2 cup	honey	125 mL
1/4 cup	brown sugar firmly-packed	60 mL
3 tbsp	vegetable oil	45 mL
1-1/2 tsp	cinnamon	7 mL
1-1/2 cups	oats quick-cooking	375 mL
2 cups	Saskatoon berries	500 mL

Preheat oven to 350°F (175°C). Lightly grease a 9 x 9-inch (22 x 22 cm) baking pan. In a medium saucepan, combine honey, brown sugar, oil and cinnamon. Bring to a boil, and boil for two minutes. Do not stir. In a large mixing bowl, combine oats and Saskatoon berries. Stir in honey mixture until thoroughly blended. Spread into prepared pan, gently pressing mixture flat. Bake until lightly browned, about 40 minutes. Cool completely in the pan on a wire rack. Cut into bars

Saskatoon Pancakes

SERVES 2 – 4

1	egg	1
1 cup	buttermilk	250 mL
1 tbsp	oil	15 mL
1 tbsp	honey	15 mL
1 cup	flour	250 mL
2 tsp	baking powder	10 mL
1/2 tsp	baking soda	2 mL
1/2 cup	Saskatoon berries	125 mL

Mix together the egg, buttermilk, oil and honey. In a separate bowl combine the dry ingredients with the Saskatoon berries. Add to liquid ingredients and mix just until it forms a batter. For each pancake drop 1/3 cup (75 mL) of batter onto hot griddle. Flip as soon as bubbles form.

Warm Saskatoon Pancake Sauce

SERVES 4

1/4 cup	sugar	60 mL
1 tbsp	flour	15 mL
	pinch of salt	
1 cup	water	250 mL
1 tsp	lemon juice fresh	5 mL
1 cup	Saskatoon berries	250 mL
1-1/2 tbsp	unsalted butter	7 mL
1/4 tsp	cinnamon	1 mL

In a saucepan, combine sugar, flour, water, salt and lemon juice. Cook until mixture thickens slightly. Add the Saskatoon berries and cook over moderate heat, stirring for one minute. Remove from heat, add butter and stir until melted. Add cinnamon and stir. Spoon sauce over pancakes.

Saskatoon Berry Torte Delight

In large bowl, cut butter and cream cheese into flour and chokecherry powder till mixture resembles coarse crumbs (with the addition of the chokecherry powder, the dough will have a light to medium purplish-red color). Gather dough into a ball. Divide dough in four equal parts. Form in balls. Wrap each separately in clear plastic wrap and refrigerate at least one hour or until ready to use. Each ball of dough will make pastry equivalent to 1 two-crust pie. This will give you 4 two-crust pies, 8 single pie shells or 20 to 24 tart shells.

Saskatoon Berry Whipped Cream
Fold the jam into the whipped cream until mixture is uniform. Divide pastry in three equal parts. On un-greased baking sheets roll each into a circle 1/8 inch thick. Using pastry wheel, trim to 8-inch circles, prick with fork. Bake in 375°F (190°C) for 11 to 12 minutes or until lightly browned. Cool thoroughly. Combine yogurt and sugar; mix well. Fold in topping and fruit. Layer pastry and filling. Garnish with additional whipped cream and whole Saskatoon berries. Chill 30–45 minutes. Cut in wedges with an electric knife.

SERVES 8

1	**recipe basic pastry** double crust	1
1 cup	**Saskatoon berry yogurt**	250 mL
1/4 cup	**sugar**	60 mL
1 cup	**Saskatoon berry whipped cream**	250 mL
2 cups	**Saskatoon berries** fresh or frozen	500 mL
1 cup	**can crushed pineapple** well drained	250 mL

BASIC PASTRY

2 cups	**butter**	500 mL
1-1/2 cups	**cream cheese**	350 mL
5 cups	**flour**	1250 mL
2 tbsp	**chokecherry powder**	30 mL

SASKATOON BERRY WHIPPED CREAM

1 cup	**whipping cream**	250 mL
2 tbsp	**Saskatoon berry jam**	30 mL

Photo Credits

It isn't without the help and encouragement of special organizations and friends that one is able to even begin the publication of a book. I want to recognize all of those very special people. Thank you.

British Columbia Blueberry Council
P.O. Box 8000-730
Abbotsford, British Columbia
Page 17, 21, 22, 24, 30, 32, 67

North American Blueberry Council
4995 Golden Foothill Parkway, Suite 2
El Dorado Hills, California
Page 6–7, 8, 10, 12, 16, 20, 27, 78, back cover (inset)

California Strawberry Commission
P.O. 269 Watsonville, California
Page 37, 40, 44, 48, 52, 54, 56, 58, 62

Driscoll's Strawberry Associates Inc.
345 Westridge Drive,
Watsonville, California
Page 34–35, 64–65, 76

Betty Hemstad
Edina, Minnesota
Page 7 (inset), 35 (inset)

Wendy L. Schmidt
Lebanon, Missouri
Page 65 (inset)

Richard Shipman
Moose Jaw, Saskatchewan
Page 80–81, 87

Rhonda Purdy
Keeler, Saskatchewan
Page 81 (inset)

Index

A
Anthony Pace's Butterscotch Blackberries, 68
Ashley Inn Lemon Curd Waffles, 11

B
Beverages
- Berry Cool Margaritas, 66
- Berry Blue Smoothie, 9
- Berry Sparkler, 9
- Berry Super Soda's, 69
- Blackberry Cordial, 68
- Blackberry Black, 66
- Blackberry Eggnog, 68
- Blackberry Smoothie, 70
- Blackberry Summer Tea, 70
- Blueberry Smoothie, 9
- Blueberry Fruit Shake, 11
- Blue Witch's Brew, 9
- Savory Strawberries on Lime Ice, 39
- Strawberry Banana, 38
- Strawberry Colanda, 38
- Strawberry Cream, 36
- Strawberry Daiquiri, 39
- Strawberry Fool, 39
- Strawberry Malted Milk Shake, 36
- Strawberry Sangria Ice, 36
- Strawberry Smoothies, 38
- Strawberry Wine Punch, 38
- Wild Berry Shake, 66

Beef Tenderloin with Blackberry Port Wine, 73
Berry Blue Frozen Dessert, 30
Berry Empanadas, 77
Berry Super Sodas, 69
Berry Torte, 78
Blackberry Cake, 77
Blackberry Cordial, 68
Blackberry Eggnog, 68
Blackberry Pie, 79
Blackberry Sauce, 73
Blackberry Smoothie, 70
Blackberry Soup – Oregon, 71
Blackberry Soup – Windsong, 71
Blackberry Summer Salad, 70
Blackberry Summer Tea, 70
Blueberry Balsamic Vinegar, 26
Blueberry Burgundy Soup, 18
Blueberry Cheese Strata, 14
Blueberry Fluffy Pancake, 13
Blueberry Ginger Sauce, 11
Blueberry Lasagna with Hazelnut Cream Sauce, 23
Blueberry Lemon Parfait, 31
Blueberry Mediterranean Salad, 17
Blueberry Melon Salad, 15
Blueberry Mincemeat, 32
Blueberry-Onion Sauced Pork Tenderloin, 20
Blueberry Orange Sauce, 18
Blueberry Orange Soup, 18
Blueberry Roasted Corn Salsa, 32
Blueberry Smoothie, 9
Blueberry Spiced, 25
Blueberry – Stuffed French Toast, 10
Blueberry Tortilla Pizza, 33
Blueberry Vinaigrette, 26
Blueberry Wild Rice Salad, 33
Blue Witch's Brew, 9
Blueberries with Creamy Banana Sauce, 26
Butterscotch Blackberries, 68

Index

C

Cakes
- Blackberry Cake, 77
- Cheese Cake-Light and Luscious, 49
- Chocolate Strawberry Ice Cream Cake, 63
- Chocolate Strawberry Port Cake, 57
- Grandma's Blackberry Cake, 77
- Little Chocolate Spice Cake, 60
- Strawberry Cheese Cakes, 49
- Strawberry Hot Fudge Sundae Cake, 55
- Strawberry Short Cake, 59

Calypso Strawberry Mango Salsa, 42
California Skinny Dips, 45
Charcoal Grilled Cornish Hens, 84
Cheesecake – Light and Luscious with Strawberries, 49
Cheese Mold – Strawberry, 43
Chef Cushman's Strawberry Chicken Salad, 47
Chef Revsin's Angelhair Pasta with Strawberries, 51
Chicken Grill with Tangy Saskatoon Sauce, 86
Chicken with Saskatoon Berry Sauce, 83
Chilled Czech Blueberry Soup, 19
Chef Martha Crawford's Double Berry Super Sodas, 69
Chocolate Strawberry Ice Cream Cake, 63
Chocolate Strawberry Port Cake, 57
Chutney-Strawberry, Taste of Summer, 46
Cinnamon Glazed Strawberries, 45
Classic Strawberry Pie, 55
Corn Salsa- Blueberry, 32
Creamy Smoked Turkey and Blueberry Salad, 17
Crab Wontons with Blackberry Szechuan Sauce, 75
Curried Chicken with BC Blueberries, 19

D

Drunken Fruit Salad, 47

E

Eggnog – Blackberry, 68

F

Fresh Blueberry and Lemon Parfait, 31
Fruit Pizza Lorenzi, 79
Fluffy Blueberry Pancakes, 13

G

Grandma's Blackberry Cake, 77
Grilled Chicken Breasts with Strawberries, 50

H

Hot and Sour Prawns with Blueberries, 22

L

Lakehouse Inn Blueberry Lasagna, 23
Lemon Blueberry and Chicken Salad, 16
Little Chocolate Spice Cakes with Strawberries, 60
Light and Luscious Strawberry Cheesecake, 49
Little Chocolate Spice Cakes, 60

M

Magnolia Inn French Toast, 13

Maple-Ginger Frozen Cream with Blueberries, 28

Medallions of Venison or Buffalo with Saskatoon's, 85

Mincemeat – Blueberry, 32

N

Nachos – Strawberry, 42

Napoleons – Strawberry Almond Cream, 53

Night Swan Blueberry Cheese Strata, 14

O

Orange Almond Blueberries, 31

Orange Buttered Rum Sauce on Blueberries, 25

Oregon Blackberry Soup, 71

P

Pasta
- Chef Resvin's Angelhair Pasta, 51
- Lake House Inn Blueberry Lasagna, 23

Pie
- Classic Strawberry Pie, 55
- The Ultimate Strawberry Pie, 57
- West Coast Blackberry Pie, 79

Pork
- Blueberry-Onion Sauced Pork Tenderloin, 20
- Pork Chops and Blackberry Sauce, 74

Prawns
- Hot and Sour Prawns with Blueberries, 22

R

Rabbit Hill Inn Blueberry Burgundy Soup, 18

Raspberries and Blackberries with Champagne Cheeses, 76

Royal Blue Poached Pears with Blueberry Sauce, 29

S

Salad
- Blackberry Summer Salad, 70
- Blueberry Mediterranean Salad, 17
- Creamy Smoked Turkey and Blueberries Salad, 17
- Drunken Fruit, 47
- Lemon Blueberry and Chicken Salad, 16
- Chef Cushman's Strawberry Chicken Salad with Hoisin-Sesame Dressing, 47
- West Coast Blueberry Salad, 15
- Wilted Spinach with BC Blueberries Salad, 14
- Tropical Blueberry Pineapple Jalapeño, 17

Salmon with Blackberry Hollandaise, 72

Salsa
- Strawberry Strawberry-Mango Salsa, 42
- Saskatoon Berry Salsa, 82

Saskatoon Crisp, 88

Saskatoon Prairie Berry Custard, 86

Saskatoon Granola Bars, 90

Saskatoon-Lemon Cheesecake Bars, 89

Saskatoon Pie, 89

Saskatoon Berry Whipped Cream, 91

Saskatoon Berry Relish, 88

Saskatoon Berry Salsa, 82

Saskatoon Berry Soup, 82

Saskatoon Berry Torte Delight, 91

Saskatoon Berry Yogurt, 88

Saskatoon Pancakes, 90

Index

Saskatoon Pancake Sauce, 90
Sauces
    Balsamic Red Wine, 59
    Blackberry, 73
    Blueberry with Creamy Banana Sauce, 26
    Blueberry Ginger, 11
    Blueberry Orange, 18
    Orange Butter Rum, 33
Savory Strawberries on Lime Ice, 39
Soup
    Blackberry, 71
    Blueberry Orange, 18
    Chilled Czech Blueberry, 19
    Cold Saskatoon Berry, 82
    Rabbit Hill Inn Blueberry Burgundy, 18
    The Colony Hotel Famous Strawberry Soup, 46
    Oregon Blackberry, 71
Scrumptious Berry Torte, 78
Spiced Blueberries, 25
Strawberries in Balsamic Red Wine Sauce, 59
Strawberry-Almond Cream Napoleons, 53
Strawberry Banana, 38
Strawberry Cheese Cake with Fresh Strawberry Sauce, 49
Strawberry Chicken Salad with Hoisin-Sesame Dressing, 47
Strawberry – Cinnamon Glazed, 45
Strawberry Cheese Tarts, 41
Strawberry Chutney, 46
Strawberry Club Sandwiches, 43
Strawberry Colanda, 38
Strawberry Cream, 36
Strawberry Daiquiri, 39
Strawberry Fool, 39
Strawberry Hot Fudge Sundae Cake, 55
Strawberry Malted Milk Shake, 36
Strawberry Mango Salsa, 42
Strawberry Nachos, 42
Strawberry Pops, 63

Strawberry Ricotta Dip, 51
Strawberry Sangria Ice, 36
Strawberry Sauté, 59
Strawberry Shortcake, 59
Strawberry Smoothies, 38
Strawberry Soup, 46
Strawberry Tart, 61
Strawberry Whipped Cream, 50
Strawberry Wine Punch, 38
Strawberry Cheese Tarts, 41
Strawberry Three Cheese Mold, 43
Strawberry Whipped Cream, 50
Summertime Strawberry Antipasto, 41

T

Taste of Summer Strawberry Chutney, 46
Tea – Blackberry Summer, 70
Tropical Blueberry Pineapple and Jalapeño Salad, 17
The Ultimate Strawberry Pie, 57
The Colony Hotel's Famous Strawberry Soup, 46
The Inn on The Common Wild Blackberry Cordial, 68

V

Veal Medallions with Blueberry Citrus Sauce, 21
Venison or Buffalo with Saskatoon Berry Sauce, 85

W

Waffles
    Ashley Inn Lemon Curd Waffles, 11
West Coast Blackberry Pie, 79
Wilted Spinach with BC Blueberries, 14
Wild Berry Shake, 66
Windsong Mountain Inn Blackberry Soup, 71